DR. DISASTER'S
GUIDE TO
SURVIVING
EVERYTHING

Essential Advice for Any Situation
Life Throws Your Way

JOHN E. TORRES, MD
NBC Senior Medical Correspondent
Col., U.S. Air Force, Ret.

HOUGHTON MIFFLIN HARCOURT
New York Boston 2021

For information about permission to reproduce selections from this book, write
to trade.permissions@hmhco.com or to Permissions, Houghton Mifflin Harcourt
Publishing Company, 3 Park Avenue, 19th Floor, New York, New York 10016.

hmhbooks.com

Library of Congress Cataloging-in-Publication Data is available.

ISBN 978-0-358-49480-5

Book design by Alison Lew / Vertigo Design NYC

Printed in the United States of America
DOC 10 9 8 7 6 5 4 3 2 1

This book presents, among other things, the research and ideas of its author.
It is not intended to be a substitute for consultation with a professional healthcare practitioner. Consult with your healthcare practitioner before starting any
diet or other medical regimen. The publisher and the author disclaim responsibility for any adverse effects resulting directly or indirectly from information
contained in this book.

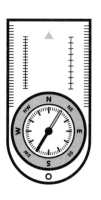

CONTENTS

Introduction   8

PART I   SECURE YOUR CASTLE   13

HOME AND AWAY   14

YOUR FIRST-AID KIT   19

Your Prospective First-Aid Kit   20

TREATING BASIC EMERGENCIES   21

Cuts/Abrasions   23

Concussion   24

Burns   25

Insect Stings   26

How to Use an EpiPen   27

Insect Bites   28

Multiple Bloody Wounds   29

Tourniquets   29

Impalement   31

Eye Burns or Splashes   32

Cardiopulmonary Resuscitation (CPR)   32

CPR on Infants and Small Children   34

Choking   35

Cardiac Arrest/Heart Attack   36

Drowning   38

Seizures   39

DECISION TREE CHART: Which Is Which?
(Stroke, Seizure, Heart Attack, Cardiac Arrest)   40

Stroke   42

THE FINE ART OF ADVENTURE PACKING   44

Day Hike Pack List   45

Going Full MacGyver   46

If You Get Lost in the Woods   47

International Morse Code   48

THE RULE OF THREE   49

HOW TO EVACUATE   51

YOUR BUG-OUT "BAG"   53

Your Prospective Bug-Out "Bag" Checklist   55

Cheat Sheet: Shopping for Your Bug-Out Bag   61

HOW TO USE A FIRE STARTER   63

HOW TO USE YOUR WRISTWATCH TO FIND YOUR WAY   66

Choosing Your Survival Watch   70

HOW TO DRINK DIRTY WATER   73

KNOW YOUR TOXIC PLANTS   76

YOUR FAMILY'S RENDEZVOUS PLAN   84

ZOMBIES   87

PART II **NATURAL PERILS**   89

ANIMAL ATTACKS   90

Rabid Animals   91

Cats and Dogs   92

Rats   93

Bears   94

Snakes   98

WILDFIRE   100

EARTHQUAKES   104

FLOODS   107

MUDSLIDES   111

Mudslide or Flood? The Great Molasses Disaster   115

BLIZZARDS   116

Surviving a Blizzard Even If You Have Shelter   117

Surviving a Blizzard in Your Car   119

Understanding Hypothermia and Frostbite   121

Surviving Hypothermia   123

Surviving Frostbite   124

AVALANCHES   126

TORNADOES   130

LIGHTNING   135

BLACKOUTS   140

Surviving a Blackout in Your Home   141

Surviving a Blackout in the Big City   144

Surviving a Blackout on Trips Abroad   145

PANDEMICS   147

Medical Preparedness   150

Special Considerations for Babies and Elders   151

Medico-Legal Issues   152

Healthy Living Choices   152

Tools, Supplies, Kits   152

Stocking Your Kitchen and Pantry   153

Your Quality-of-Life Needs   154

Your Personal Protective Equipment ("PPE")
Disinfecting Station   155

Traveling During a Pandemic   157

PART III HUMAN-CAUSED DISASTERS   159

GAS LEAKS   160

Natural Gas Tips   161

Carbon Monoxide   162

When Hospitals Are Necessary   163

Specialized Gases   163

HOME OR BUILDING FIRES   165

Surviving a Fire in Your Home   169

Surviving a Fire in an Office or
Commercial Building   171

TERRORIST ACTIVITIES   173

ELEVATOR ACCIDENTS   179

BIOTERRORISM   181

    Surviving Bioterrorism   185

CHEMICAL AGENTS   188

    Surviving a Chemical Attack   190

DIRTY BOMBS   193

    Surviving a Dirty Bomb   194

SHOOTERS/STABBERS   196

    Tips for Gun or Knife Attacks   199

CIVIL UNREST   201

    Surviving Civil Unrest   202

CYBERATTACKS   206

    Good Web Hygiene   207

    Spam Calls   209

    Password Protection   210

    Dealing with Companies Online   211

    Personal Data   212

    Loved Ones and Scams   212

    Securing Medical Data   213

AIRPLANE DISASTERS   214

TAKE YOUR PULSE   218

Acknowledgments   220

Disaster Cheat Sheets   222

Resources and Where to Buy Gear   232

Index   236

About the Author   240

INTRODUCTION

The book you're holding in your hands should never be used as a flotation device. And you'd be foolish to try to use it to fend off a bear attack. But despite these few shortcomings, I am quite certain that its contents could someday save your life or the life of someone you love.

No one likes to think about disasters. But they occur every single day: Earthquakes in Italy, California, or even North Carolina. Tsunamis in Thailand, Japan, or elsewhere in the Pacific Rim. Massive flooding everywhere in the world, from the American South to northern England. Dam failures and massive mudslides in Brazil. And when the disasters are not due to our dynamic planet and the natural world, we have disaster wreaked by human hands: Active shooter incidents in American elementary schools or on a high-speed train bound for Paris. In recent years we've had two terrifying stabbing attacks on London Bridge alone. Pedestrian ramming attacks and bombings in Barcelona. Now-famous chemical attacks on innocent civilians aboard Japanese subways. Civil unrest in Belarus and in every major city in the United States.

No one wants to envision scenarios in which bad things happen to them. Indeed, part of our unconquerable human spirit practically demands that we look away and try to convince ourselves that It Can't Happen Here. But since the COVID-19 pandemic, all of us are a little closer to becoming "preppers"— folks planning for that inevitable day when everything goes wrong.

It doesn't really matter what part of the world you live in, or where you just happen to be visiting. Natural and human-caused disasters happen all over the world. I think it pays to understand why disasters happen, how you can keep your family safe, and what you can possibly do to survive what fate slings your way.

Folks, let me just tell you: Disaster is my frenemy.

Most people know me as NBC's senior medical correspondent. But before I was a "TV news doctor," I was an emergency room physician and trauma specialist. (And still am.) Before that, I was in the National Guard. Before that, I was an Air Force pilot.

During Operation Iraqi Freedom, I was stationed in Balad, Iraq, and served as a flight surgeon. I served in rescue missions to Antarctica, treated victims of Hurricane Katrina (New Orleans, 2005) and Hurricane Rita (U.S. southeast, 2005), worked in diplomatic missions to Jordan to develop that nation's emergency room system, and treated children in need on numerous humanitarian missions to Central and South America.

I still serve as an instructor for NATO Special Forces. I teach soldiers, medics, and doctors a variety of subjects. Tactical combat casualty care. Combat simulations. Medical leadership. And the big one: bioterrorism.

Some people call me a Disaster Geek. Well, guilty as charged. You could say that in all the roles I've ever had, it's been my job to obsess about what can go wrong. And boy, have I.

I first learned survival skills from my father, who took me camping and hunting as a boy in Europe and the United States. My father was an Air Force security policeman who also happened to be an amazing outdoorsman. He could track any animal and hunt it down. He could find his way out of a wilderness situation with a compass alone. He could build shelters with nothing but a knife and some cast-off vegetation.

I remember one time joining him on a hunt in the New Mexico wilderness. Each of us was trying his luck in neighboring canyons. I was striking out big time, and was about to throw in the towel when I heard a gunshot ring out. My dad had probably fired at something. But had he hit it?

It took me a while to hike out of "my" canyon. It was slow going, because that rough desert terrain is no joke. I was puffing as I was heading up the hill that would lead to Dad's turf. Only 20 minutes had elapsed since I'd heard the gunshot, but there was my dad—this five-foot-nine-inch-tall Latino man, heading in my direction down a hill, dragging after him an enormous elk whose entrails he'd already cleaned out to keep the meat from spoiling. That was important because we never hunted for trophies. We hunted for food.

Only a few months into the Air Force Academy, I was on the phone telling my dad that I didn't think the Air Force was for me. I was thinking of quitting.

"Do me a favor," Dad told me. "You have that SERE course coming up soon, don't you? Don't quit until you take that."

SERE training—which stands for Survival, Evasion, Resistance, Escape— is one of the most grueling and demanding courses military personnel can take. Your trainers leave you stranded in the wilderness with nothing but a knife, a compass, a map, and a minuscule amount of food. You pretend that you've dropped down behind enemy lines. Your job? Stay alive, and get back

to your unit. Later, they try to break your will in a simulated prisoner of war (POW) camp. Your job? Don't talk, and figure out a way to escape.

Those first few days into the wilderness scenario, you feel like you're going to die. But then you wise up. You learn that drinking muddy water isn't delicious, but will actually quench your thirst. You learn how to snack on insects for their protein content. You learn how to live off of every part of every animal you capture and butcher. When my compatriots and I got hold of a rabbit, we were ecstatic—but I was the one who got stuck gobbling up the bits no one else wanted. You don't want to know what a rabbit's eyeball tastes like.

You know what? My dad was right. After SERE training, I didn't dare quit the Air Force. Those 3 weeks of grappling with potential disaster had given me a new mindset, an unquenchable attitude that has lasted to this day:

If I can do this, I can do anything.

This book is an armchair course on cultivating that mindset. Nine times out of ten, the best way to handle a disaster is to not panic and work practically to solve the problem.

But there are two other important lessons I want to impart before we get into the meat of this book.

The U.S. Air Force Pocket Survival Handbook shares two shocking cautionary tales about two different men who were stuck in the middle of nowhere. One man was stranded in the Arizona desert and made it back alive, surviving 8 days without food and water. The other was a pilot who crash-landed on a frozen lake in the Canadian wilderness and killed himself only 24 hours before a search-and-rescue team found him. Why did one man survive while the other gave up so quickly? Clearly, one had *the will to survive,* the other did not. I am sure you would find the same among many survival stories. Avalanche experts die in sheets of snow. Lifeguards drown. I can't help thinking that the ones who make it fought harder because they were trying to get back to the people they loved.

In the military they always pounded into our heads the following directive: "Keep your head on a swivel." In other words, always be aware of your surroundings. Know what's in front of you, what's in back of you, what could potentially come at you from the side. Until God starts making humans with eyes in the backs of our heads and over our ears, a swiveling head is your best tool for staying alive. This lesson is worth repeating a million times because too often these days we're distracted by our mobile devices. Whether the disaster you're facing is natural or human-caused, you must anticipate it coming, and know what you're going to do if it does.

Summing up, let me state the two big lessons you will hear me repeat in a multitude of different ways in this book. *Stoke your will to survive. Keep your head on a swivel.*

STOKE YOUR WILL TO SURVIVE.
KEEP YOUR HEAD ON A SWIVEL.

There you go. Those are the two biggest concepts. You can learn the rest, the way those of us who deal with disasters have. When a bleeding patient is wheeled into an emergency room on a Friday night, the doctors on shift have no idea what they are about to deal with. But they have three strengths in their corner:

- They know in their gut that they can handle anything that comes their way because they've seen it all before. *If I could do what I did last week, I can do anything.*

- They have mad skills learned in med school and on the job.

- And they have a ton of great tech at their disposal.

In this book, I will show you how to assemble the tech you need, and I hope to instill some of the same confidence.

But right now, I'll let you in on a secret: You don't really need to have a doctor's skills to save the day. For example, a civilian doesn't really need to know how to, say, stitch a cut in someone's scalp. All they need to know is that strips of duct tape will keep a wound closed enough until they get their loved one or pal to the emergency room.

In other words, harnessing your inner MacGyver is easy so long as you cultivate the mindset (*I can deal with this. I can deal with anything, so long as I don't panic.*) and have the tools (duct tape).

This book is divided into three parts—survival basics, natural perils, and human-caused disasters. Along the way, you'll find some inspiring stories, short sections with tips, checklists to help you assemble your critical gear, and occasional drawings to shed light on the tricky stuff. I have tried to avoid mentioning the brand names of products unless absolutely necessary to get my point across. You will find a list of products and gear to get you started in the resources section. Keep in mind that these are merely suggestions, not hard-and-fast recommendations. Also, every person's needs are different.

Read and enjoy this book any way you wish. Dip in and out of sections as you like. Or just familiarize yourself with the content, keep the book handy, and grab it when you need it most. We made it small enough to take on the road or stuff in a backpack for that very reason.

And don't worry: I'm not going to make you eat insects.

But I will share one profound piece of insect wisdom culled from a story written 2,500 years ago by a fellow named Aesop.

It comes down to this: In the world of ants and grasshoppers, you want to be an ant. You want to be prepared for whatever life slings your way. And once you are, you can help any and all grasshoppers who need the help.

I've got your back, so you've got theirs.

DR. JOHN TORRES
Colorado
September 2020

PART I

SECURE YOUR YOUR CASTLE

HOME AND AWAY

When I was working in the ER, I got really good at sizing up people I saw in the waiting room. One perennial favorite was the family hanging out together after a weekend mishap. Two members of the family—usually Dad and one of the children—had ice packs clapped to their foreheads, nursing swollen bumps on their heads. In addition to the ice-pack, the boy or girl had an arm or leg that needed to be looked at. And Mom always looked pretty angry.

I'd see these folks and my mind would immediately think: "Trampoline accident."

And often I was right.

Trampolines are totally fun, and I do not want to scare you off buying one for your family. But I could never bring myself to do so. Here's why: When one child (or one grown-up) hops on a trampoline, they will most often stay safe because one human on a trampoline will tend to bounce up and down in the middle of the mat. That's the bounciest part, and the safest because you remain far from the edges. But when two people hop on a trampoline together, they sort of move away from the center and rove around all parts of the mat. If two little kids are having fun together, they're usually okay because they tend to weigh about the same, so no one child's hop is going to be particularly unsettling for the other. But inevitably, when Dad decides to hop on the trampoline with one or all of the children, Dad doesn't know his own strength—er, weight. The hops become asynchronous. Dad's one giant hop throws a kid off balance, the kid goes flying, and the unhappy result is the family unit in the waiting room. The family that hops together gets casts together.

Yes, trampolines today are made with safety nets, but those nets are imperfect. They're usually very loose. Sometimes they tear and replacing them is a pain. But even if they're in good shape, they aren't really strong enough to stop a falling body.

Trust me. I've seen enough cases in the ER to know.

I guess you could say that the military life, combined with a career in medicine, has wired me to be a little like a perennial Boy Scout who is forever living the first rule: Be Prepared.

By now my wife Lynda and my kids have come to accept my strange little habits. Lynda knows that I sleep with shoes near my bed. My kids know that Dad keeps 30 days of medication—over-the-counter drugs and any prescriptions any of us are currently using—in the house at all times. (Yes, you can ask your doctor to prescribe in such a way to accommodate this approach for most medications.) My family knows that as soon as our stockpile drops below 30 days, I reorder whatever we need.

Because. Just because.

They know that if you dropped me in any city, town, or rural area in the United States, I'd need about 2 hours to visit various big box stores and drugstores to assemble a solid first-aid kit. The resulting go-bag might contain such things as bandages, malleable splints, a box of nitrile plastic gloves, a Leatherman or Gerber multitool, a very dorky headlamp flashlight, duct tape, enough fabric to fashion several tourniquets, iodine tablets, a bottle of bleach, a gallon of water for every person in the family for however many days we need, a fire starter or pocket flint, and a box of disposable N95 dust masks to protect ourselves from inhaling dangerous dust and contaminants.

I'd combine these essentials with a multitude of other things it always makes sense to travel with—like ROAD iD–brand silicone bracelets emblazoned with every family member's critical information, such as name, blood type, allergies, and emergency contact information.

If I had just a little more prep time, I could probably arrange to order some LifeStraws, which would allow me or my family members to drink almost any type of standing water safely.

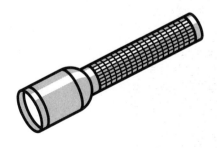

My kids used to think their old man was a little nuts to go to such extremes. But today they are both doctors, and I occasionally hear them espousing things I've said along the way. They don't always credit me. They don't have to. As long as they take precautions, I'm happy. They're old enough now to think for themselves. But when they were younger, my mind worked the way it did so theirs didn't have to. I think that's what you want for your family and for everyone who lives under your roof—pets and grandparents included.

Your home is your castle. There's no place like home. Home is where the heart is. We use these kinds of sayings to show the special regard in which we hold the place that shelters us. These days, I think it really makes sense to take the time to ensure that your home is safe, and that you have taken some simple precautions. A lot of the stuff I'll mention in Part I are things you may have heard all your life. But I'm putting them here, all in one place, so you have it when you need it.

Let me suggest that you challenge yourself to work through each of the sections and nail down what makes sense for you and your family. Get your home life squared away first, goes the thinking, and then you'll be ready to tackle what happens when you leave the house.

To that end, here are just a few action steps you might want take around the house, maybe as soon as the coming weekend. I consider them to be the low-hanging fruit of home safety, designed to nip many of the classic disasters before they happen. You'll find more as you flip through the rest of Part I.

- We all have annoying little things around the house that require attention. Uneven walkways, doors that don't latch properly, outdated electrical outlets that need to be upgraded, old appliances that give us trouble, etc. Dealing with these is onerous because they require time, money, and often energy to track down a suitable contractor. Make it easy on yourself: Choose a couple of projects to deal with per each quarter of the year, and march through them as time allows.

- Make sure that questionable tree limbs are trimmed away from power lines. Make sure that garden tools, snow shovels, etc., are in good condition so you can take care of your yard across several seasons.

- Check all the ladders and stepladders you have, and replace any that seem rickety or in bad shape.

Gas leaks have sent me running out of my home after midnight at least twice in my life. As a result, I always leave a solid pair of slip-on shoes near my bed when I go to bed at night. Not slippers. Good-quality shoes with robust soles that will safely carry me outdoors in case a gas leak, not to mention a fire, or any other emergency forces me and my family to flee our home in the middle of the night. If that makes me paranoid, so be it. Go and do likewise.

Be certain that your garage door operates properly. If you have young children, consider an electric door opener. Be sure the "electric eye" that stops and reverses the door if it detects an obstruction is operating correctly and free of debris like leaves, cobwebs, etc. See that the springs and cables of the door are lubricated periodically, according to the manufacturer's instructions. Be sure that the backup battery on the opener is fresh. You want to be able to access this space and all that you may be storing in this room in the event of a power outage. You'll also want to be able to retrieve your vehicle quickly if an evacuation is necessary.

Once a week, make the effort to drive your car so the battery stays charged, and try to keep the gas tank topped off so you can leave town without having to stop for gas. It's smart to keep a rescue tool in the vehicle so you and your passengers can escape in case the vehicle becomes submerged. These combo-tools have a blade to slice through seat belts, and a hammer that can shatter the tempered glass in side windows, but not windshields. (Please note: Gasoline goes bad after 6 months. If you don't drive your car for a long period of time, you may be forced to siphon out the bad gas, and replace it with fresh fuel.)

If you are planning to install a chest or upright freezer in your garage, be sure that you are looking for "garage-ready models" that are designed to withstand the temperature extremes of an unconditioned space.

Your smartphone probably has an app that will allow you to select your emergency contacts and show them on the screen of your locked phone so first responders can easily see that information if something happens to you. Enable this feature now. Your phone may also be able to automatically send your location and vital statistics when you dial 911—another good feature to enable. As a last resort, select your emergency contacts in your address book, and type "ICE"—"in case of emergency"—after their names. Just understand that if you go this route, in an actual emergency first responders may not be able to unlock your phone to see who to call.

When our pets leave the house, they carry contact information on their collars so we can be reached if the pet goes missing. But every day we send our children out into the world without any ID. That's reasonable, I suppose, if you're concerned about strangers knowing your child's name. But you might consider rethinking this position while on family vacations, or while visiting crowded locations like theme parks where your kid can easily get lost. A secure, comfortable bracelet (i.e., ROAD iD) that lists the child's allergies, known medical conditions, and your contact information is a good choice while on the road.

Seasonally inspect all the sporting or recreational equipment your family uses (lifejackets for water sports, bike helmets, etc.) and replace those that are damaged or that your children have outgrown. As you'll see throughout this book, sporting helmets—such as bike helmets or football helmets—can be used in a pinch to protect your skull in certain emergencies.

If you're looking for ways to teach your children about disaster preparedness without alarming them, check my section on zombies (page 87.) Yes, I said zombies.

While you're at it, humor me. Check the safety net on the trampoline. You're welcome.

YOUR FIRST-AID KIT

E very home should have a good first-aid kit. You can easily find a well-stocked kit online or in a local drugstore, but be aware that pre-assembled kits will not have every single supply you need for every single situation. It's fine to buy a pre-made kit, but once you get it home and familiarize yourself with its contents, take the time to amplify and customize the kit to meet your family's needs. Sometimes you can anticipate what kinds of crazy scrapes you or your family are going to get into.

A seasoned home chef, for example, can assume that there will be some burns and minor knife cuts on the way to culinary genius, so they might wisely beef up their store-bought kit with some additional gauze, cold packs, and burn pads. Families with small children will want everything they need to treat minor cuts and bruises, and perhaps a good supply of child-friendly cold medication, ibuprofen, and so forth. Both of these families might well have a few flexible cold pads chilling in the freezer.

Store your first-aid kit where it's easily accessible. If you have numerous rooms or levels, it pays to store more than one kit on different floors or rooms of your home, so you don't have to run far to retrieve supplies in an emergency.

The following list is far from comprehensive, but it will get you started and thinking about your personal situation.

Your Prospective First-Aid Kit

☐ Analgesics (i.e., acetaminophen, aspirin [adults only], ibuprofen, naproxen, etc.)

☐ Antibiotic ointments (i.e., Bacitracin, Neosporin, etc.)

☐ Antiseptics (spray, lotion, wipes for cleaning wounds)

☐ Bandages (various sizes of adhesive bandages, rolls of gauze, gauze pads for larger wounds)

☐ Cold packs

☐ Elastic bandage (i.e., ACE)

☐ Flexible aluminum splints (i.e., SAM Splint)

☐ Flexible cold pad (i.e., Thermipaq, stored in family freezer)

☐ Hydrocortisone cream (to treat itching, swelling)

☐ Plastic disposable gloves

☐ Scissors

☐ Self-adherent wrap (i.e., Coban)

☐ Thermometer

☐ Tweezers (pointed-tip version)

☐ Waterproof adhesive tape

TREATING BASIC EMERGENCIES

The human body is a remarkable organism. But it is also a strange study in contrasts and unpredictability. A person can be fit as a fiddle, capable of running a 5-minute mile, and still trip on an uneven walkway visiting a relative and end up in a cast. Minuscule organisms reside in the human gut and assist us with our digestion, but a tiny prick on the hand from an unclean pocketknife will introduce bacteria—and a painful infection—that may necessitate amputation. The human stomach will happily digest the bloodiest of steaks, but if any of your own blood ends up in your stomach, your body will violently reject it and vomit it up. A person can live their whole lives with a foreign object—a bullet, shrapnel, a titanium surgical rod—lodged somewhere in their system, but patients who receive donor organs need drugs for the rest of their lives to keep from rejecting what the body will always perceive as "foreign" tissue.

Doctors spend 10 years of their lives training to comprehend all the complexities of this astounding living machine. But we can't be everywhere. Emergencies happen in the real world. That means that before a medic or ER doctor lays eyes on a patient, the person was helped or assisted by medics, loved ones, or Good Samaritans who kept calm, called for help, and did the right thing to keep the person from getting worse.

What's the right thing?

It's impossible to say because every case is different. But if you look at it another way, sometimes it's dead simple.

I'm a fan of the TV show *Game of Thrones*, which was a huge hit for 8 years on HBO. It's your basic medieval fantasy drama, a very "adult" show, complete with swords, dragons, and tons of violence, based on a series of bestselling books by George R. R. Martin. Early in the series and books, one of our young heroines, Arya, meets a man named Syrio Forel, who becomes her first mentor, teaching her how to fight with a sword.

Forel loves spouting juicy words of wisdom. On Day One of Arya's training, he says, "All men are made of water, do you know this? When you pierce them, the water leaks out and they die."

In a nutshell Forel has expressed one of the basic rules of human survival. Of course, he's not really talking about water. The job of a swordsman is to end life. The job of a doctor is to save it. And sometimes, in the case of an emergency, it becomes your job too. If that happens, you need to keep in mind some basic rules.

ONE: Stop the bleeding.

TWO: Your hands are weapons for good. Use them. Applying direct pressure to wounds can save a life.

THREE: All humans need oxygen. Help the injured person breathe.

FOUR: Call for help, or ask someone nearby to call 911, or the emergency services number in whatever country you're in. (See the link on page 235 for emergency numbers in other nations.)

FIVE: Know and master the skills you need to help people in your immediate family, especially those with special medical concerns. If your child has allergies, become an EpiPen master.

SIX: Everyone could benefit from learning cardiopulmonary resuscitation (CPR) and the Heimlich maneuver to prevent choking. If you cannot commit to a class, read up on these two skills or watch some videos.

In the section that follows, I'll show you how to treat some very basic emergencies that you might encounter in the home or in the wild. As you read through these, try to appreciate the underlying principles at work. I think you will find that a lot of it boils down to simple things such as clean the wound, don't make things worse, and, for heaven's sake, don't let the precious "water" leak out.

Cuts/Abrasions

This is the most common wound you'll encounter in the home, while playing sports, working around the house, or gardening.

For a small cut or scrape, the best thing to do is the most obvious. Wash the wound, pat it dry, and apply a bandage.

Avoid using a cleanser such as hydrogen peroxide. People tend not to want to touch or disturb broken skin because they think they'll cause the patient more pain. So they use hydrogen peroxide, which feels like a touchless way to wash out a wound. It definitely works, but the solution interferes with new cellular growth, which slows healing. If you must use it, use it the first time you clean the wound, and never again.

If you're outdoors, use drinking water to clean that wound, not lake, river, or pool water. Those are all likely to introduce bacteria into the wound.

The snip of garden clippers. The rake of a kitchen mandoline across a finger. The bite of a chef's knife. Deep cuts are another thing entirely. They're often more painful, and very bloody. Direct pressure is the best way to stanch the flow of blood. Wash the wound under running water, clamp your hand directly down on the opening, and tighten paper towels or gauze on the spot until the blood stops.

If the cut is very deep, you may encounter the situation in which the blood stubbornly refuses to stop. Get the person to a doctor. If the blood keeps coming through the gauze, apply a second layer. And a third. And a fourth...never remove the gauze. If you do that, you'll tear the wound open again and pull off any life-saving scabs that may have formed in the meantime. If you're destined for an ER, doctors will have access to tools and techniques—sutures, cauterization, powders that form artificial clots—that no mere mortal keeps at home.

Concussion

There are tons of misconceptions about concussion. People seem to think that a person has to pass out—literally hit the deck and be unconscious—in order to have suffered a concussion. Not true. If someone suffers a bump on the head, they might very well remain standing but exhibit some strange behavior. Your job is to be on the lookout for those changes.

A person who displays any of these "stroke-like" symptoms should be transported as quickly as possible to an emergency room or medical facility.

- They pass out.

- Their speech is slurred.

- They complain of numbness in hands, fingers, or other extremities.

- They don't recognize people or situations around them.

- They exhibit outright amnesia or forgetful behavior that was not present before the incident.

- Fluids are leaking out of their ears and nose. (Blood means they've damaged capillaries in their head. Clear fluid from the ears or nose could be spinal fluid, and a sign that the bump was so severe that the meninges—the delicate protective layer that covers the brain and spinal column—has been damaged.)

- Pupils of the person's eyes should look perfectly identical. Something that looks a little "off"—one pupil larger than the other, say—is not a good sign.

- Excessive or projectile vomiting.

If none of these signs are immediately apparent, great. The human head can take a fairly good thump and still be fine. You'll just get a bump that needs to be iced down. Wash off any debris or abraded skin and treat minor cuts, then use an ice pack to cool the bump. Apply a cold pack or ice for no more than 20 minutes at a time. If you're using real honest-to-goodness ice, wrap it in a towel or some other fabric first. Ice placed directly on skin will burn the flesh. Not good. (By the way, stop the bleeding, but don't be surprised by the amount of blood. Heads and face bleed a lot!)

In movies, after someone suffers a bump on the head, you'll often hear another character yell, "Don't fall asleep! Stay with me! Stay with me!"

It's actually perfectly fine for a concussed person to take a nap. But someone needs to keep an eye on them. I'm not saying you have to sit by the bed and watch them like a hawk. Just remember to check on them every 30 minutes or so. Jostle their arm and rouse them just enough to get them to answer a few questions.

How are you doing?

How do you feel?

Can you tell me the time?

You're basically trying to assess if they can think clearly. "Leave me alone! I'm trying to sleep!" is a perfectly normal response, by the way. It shows they're alert enough to know they've been disturbed.

It's normal to experience nausea, vomiting, or a bad headache. You can treat all that on your own by letting the person get some rest. But if symptoms persist or worsen—vomiting turns to blurred speech or impaired thinking or even projectile vomiting or constant vomiting—then it's time to seek medical attention.

Burns

When a burn first occurs, get the injured person out of the range of danger, remove any clothing, and clean or wipe away any chemicals, food products, or anything that may have caused the burn.

Burns aren't fun. The patient feels like the heat is building up, and the whole time that is happening, the skin continues to be damaged. The remedy is to cool the skin as quickly as possible. At home, the easiest thing to do is to hold the injured skin under cool running tap water. If you cannot do that, immersing the burn in ice water for a few minutes at a time, then waiting, then immersing again, is the best bet. Chemical cold packs can also help if you have one handy. One of the old-time remedies is to smear butter on a burn. Never do that! However delicious, butter is not sterile, and it's an organic product—made from cow's milk—so it is not stable and can be absorbed into the person's skin.

Doctors once classified burns as first, second, or third degree—indicating the degree of severity, with third as the highest. These days we say a burn is either "partial thickness" or "full thickness."

If you poke the person's burn gently and they yelp, "Ow!"—that's a good sign. It means that the nerve endings are working perfectly fine, and though the skin damage looks bad, the skin will grow back after a blister forms. If a poke at someone's burned skin elicits no response at all, they must be seen by a doctor. They may have sustained full-thickness damage. That's when the skin has died, and the person will need surgery, typically a skin graft, to protect the underlying organs.

In a partial-thickness burn, you should allow the blister to form but don't pop it. The blister acts as an enclosed, sterile environment that protects the next layer of skin. If you leave them alone, blisters eventually release fluid on their own, in their own sweet time, when the next layer of skin is ready to emerge.

Antibiotic ointment (such as bacitracin) will help clean the site. Other types of analgesic ointments (such as lidocaine) may help with pain, but don't expect miracles. Ointments only cover the top layer of skin, and the pain is deeper down. Accept that you're going to have to live with a certain amount of pain with a burn.

Cover the site with a loose gauze bandage. Tape it in place. Inspect the wound on a regular basis and change the bandage as necessary.

Insect Stings

Vespid (wasp, bee, and hornet) allergies are quite serious and life-threatening. If you're treating someone who has suffered an insect sting, ask immediately if they're allergic and if they have an EpiPen. If so, move directly to administering the pen (see "How to Use an EpiPen" on the next page).

Insect anatomy varies greatly. Hornets and wasps are capable of repeat stings. Bees can only sting once, and they leave their venom sacs behind. Remove a bee stinger as quickly as possible, ideally by scraping it out of the wound with a credit card or playing card. Only as a last resort should you pluck it out with your fingers. That's a sure way to introduce more venom to the site.

When the stinger is out, wash the wound, pat dry, apply ice. When the pain subsides, you can apply antihistamine cream, which will reduce the itching.

Be aware that some people may not know that they're allergic, either because they've never had an insect sting before, or because previous encounters were without incident. Be alert for signs of a reaction. Hives may erupt on the person's body, not necessarily in the area of the sting. They may experience trouble breathing, or throat itching or swelling. If this happens, transport immediately to the closest hospital or have someone summon an ambulance.

HOW TO USE AN EPIPEN

Anyone who has serious allergies will have a prescription for injectable epinephrine. This is a hormone that works by flooding the person's body and suppressing their overactive allergic response. This allows them to breathe better and maintain normal blood flow until they can be seen by a medical professional.

EpiPens are the best way of delivering this critical hormone when they need it. The "pen" contains a pre-loaded syringe that injects itself into the patient's body—provided you follow the correct steps.

When problems occur with EpiPens, they're usually due to people who have never administered them before. Believe me: Allergic individuals and their close family members are usually quite practiced at delivering the shot. There's usually no reason for an unpracticed bystander to do the job. But if the patient cannot do it for themselves, and no one else with experience is nearby, here's what to do:

○ Remove the pen from its plastic carrying case.

○ Remove the blue top. Keep the orange tip down. This is the business end of the syringe.

○ Grasp the EpiPen in one hand, keeping all fingers away from the orange tip.

○ Swing the pen and hit the patient in the upper thigh muscle at a right angle. You'll hear a click as the orange tip retracts and the syringe jabs through their clothing to deliver the epinephrine.

○ Leave the syringe in and count slowly to three: "One-one-thousand, two-one-thousand, three-one-thousand."

○ Remove the needle from the person's thigh. Massage the thigh, or have the patient massage it for a few seconds.

○ If you use an EpiPen on someone, call 911 or get them to medical care quickly. The epinephrine could wear off before the allergic reaction does, and they might need more.

○ Besides EpiPens, allergy sufferers may use Adrenaclicks, which operate in basically the same manner. Adrenaclicks require removal of two grey caps, top and bottom. The needle is contained in the red tip, which is inserted into the thigh for a count of 10 seconds.

Insect Bites

Mosquito bites are usually uneventful. Wash. Ice. Apply antihistamine cream or solution, and avoid scratching the site. The wound, such as it is, will subside and disappear in a few days. Occasionally these too will trigger an allergic response, so monitor accordingly. If the site becomes red and swollen, get it checked out by a doctor.

Ready to be grossed out? Great. Let's talk about ticks! Carriers of Lyme and other diseases, these minuscule creatures survive by burrowing into the flesh of mammals, and feasting on their blood. If a tick stays attached for more than 24 hours, the host's blood fills the tick's belly and becomes contaminated by various germs residing in the tick's gut. The host becomes infected when the tick regurgitates the host's blood back into the host's body. That's why keeping track of time is important.

If you spend time outdoors on a regular basis, consider using an EPA-approved tick repellent (see page 233), and wear long pants and socks when you are hiking in the woods. When you return from the great outdoors, check yourself, your partner, and your children to ensure that ticks haven't gotten under their clothing and attached themselves to their skin.

If you do discover a tick that is still attached, first take a couple of good, clear photos of the insect. You can use this image to later identify the specific tick and assess the likelihood of disease. A phone or digital camera will record the time of the discovery, which is also useful.

Using a tweezer, grab hold of the tick's head and gently pull it from the body, keeping the tool parallel to the skin. Don't pull quickly; that will almost certainly break the head off. And never grasp the tick by its swollen, blood-filled belly. That will only inject contaminated blood into the person's skin. Don't worry if the head breaks from the body and remains lodged in the person's skin. Place the remainder of the insect in a plastic bag for later inspection by an expert.

Clean the area with soap and water, apply antihistamine ointment and a cold pack, and seek medical attention. You can submit your photo for identification at various tick ID sites such as the University of Rhode Island's Tick Encounter Resource Center. Be sure to give the site operator some idea of how long the tick was on you or your loved one. Typically, they'll respond with an identification and an assessment of your risk of exposure to various diseases.

Multiple Bloody Wounds

You've been in an accident and your best friend is bleeding from their head and arm. Which do you treat first? A head wound sounds more serious, but you should always treat the wound that is losing the most blood first. Blood loss is always the most dangerous situation, and the quickest way to a fatality. Apply direct pressure and stop that blood. What doctors mean by direct pressure is not often understood by the general public. Do not be afraid to push hard. Once you have applied pressure, you should not take your hand or hands away until you have a bandage in place. If there's no one else around, you could conceivably apply pressure to two wounds at the same time, and call for help. But if you have access to fabric—a shirt, a hoodie, a pair of trousers, a belt—by all means use that to make a makeshift bandage, then move on to the less serious wounds.

Tourniquets

On the battlefield, military medics use pre-made tourniquets that are constructed out of ballistic nylon fabric and are equipped with a rigid steel or plastic rod that can be inserted and twisted to tighten the tourniquet. The twisting part is known as a "windlass." You can easily buy these types of tourniquets online, but chances are you'll never have one when you need it. It's just not something civilians travel with, because no one ever expects to treat a wound that is so severe. But you don't need a commercially built tourniquet; you just need some common sense.

Tourniquets get a bad rap because people have historically used them incorrectly and caused greater harm by cutting off circulation and causing the loss of a limb. Some principles to keep in mind: Never use string, wire,

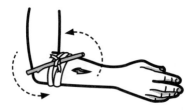

cords, or rope to fashion a tourniquet. Wide fabric is ideal. A shirt sleeve. A scarf. A pant leg. A lengthy sock. All of these would be better than rope, believe you me.

Wrap the fabric a few inches above the wound and tie a knot, leaving some extra fabric dangling off. Next, attach those dangling ends to a stick and finish a knot snugly above it. You can now twist the stick to tighten the fabric and increase the pressure on the wound. If you don't have a stick, almost any long object will work: the handle of a tennis racket, a golf club, the handle of a garden tool, or even a hammer. In a pinch, I've seen people use a couple of plastic ballpoint pens. (One pen will snap in two; a few work well.)

Once you've got your tourniquet in place, you will need to keep the windlass (i.e., the stick) from spinning around and loosening the tourniquet. If the patient has the presence of mind to do this, ask them to hold it steady while you call for help or attend other wounds. If they cannot assist you, use some other object to impede the stick from unraveling. You can loosely tie a shoelace around the arm or leg and attach it to the twisting stick.

Keep your eye on the tourniquet as you wait for help to arrive. Make sure that the fabric is not cutting into the skin. You have to do this without loosening the tourniquet once you have twisted it. You'll only break up the healing blood clots that have formed.

One last word: Belts are not ideal for making tourniquets because they are difficult to twist. Avoid them.

Impalement

The world is filled with sharp objects, and sometimes they painfully end up in the human body. A projectile that's sent flying by an explosion. A stick someone was playing with. A sharp piece of metal that snags a floating person in raging floodwaters. A wrong, painful step at a construction site. It's all possible, and I've seen it all.

A couple of rules of thumb. If the object is in the eye, leave it there. Attempting to remove it might well deflate the eyeball. Leave the job to a professional. Wrap the damaged eye with the object intact, and then wrap the second, unharmed eye as well. Here's why: If you don't completely blindfold the person, they will continue to use the good eye to look around and interact with their surroundings. Each time the good eye moves, the muscles of the bad eye will move as well. Not good. Immobilize them both by keeping them covered.

If the projectile or object is in the arm or leg, remove it if you can do so easily, apply direct pressure, bandage, and get help as quickly as possible. If the object is in the abdomen or chest cavity, leave it in, wrap the wound, and get them out of there. Trying to remove something in that area of the body risks damaging organs that haven't already been punctured.

The only exceptions: If you can't get to a hospital within the hour, or the impaling object is fixed in place and you can't take the time to cut the person free, then work the injured person free, treat the wound as you would any other serious wound, and get them to a hospital.

I knew a woman once who had fallen from the back of her truck onto a perpendicular piece of rusty rebar sticking out of the ground. The steel penetrated her buttock but missed all vital organs. She was lucky, but impaled like a bug in a display case. She was completely alone, and had no tools to remove the rebar. (That kind of steel is so tough you need a torch or diamond-tipped blade to cut it.) Her only choice was to do what she did: slowly, painfully pulled her butt up and off the top of the steel and crawled to her mobile phone to summon help. She was probably the toughest patient I ever treated in the ER. Talk about the will to live!

Eye Burns or Splashes

These types of accidents typically occur in workplaces and homes, wherever dangerous liquids are kept. The only good news about that is the proximity to professional medical attention. Don't waste time. You need to flush the injured person's eye with copious amounts of clean, lukewarm water. You will need more water than you think. A kitchen sink is ideal for this because they are often equipped with spray nozzles that you can extend, point at the person's face, and spray directly into the open eye or eyes. If the sink does not have a spray nozzle, another way to go is to fill a pitcher or other large container, and pour directly into the affected area.

Bathroom sinks are usually not the best place to perform this operation because the faucets are typically low and the basins shallow. But if that's all you have access to, stick with it! Use whatever container you have handy to scoop up water and pour it into the victim's eye.

Rinse the eye in such a way that the contaminated water goes from the eye directly into the sink, floor, or ground outdoors—not the good eye. For example, if the contaminant is in the person's left eye, have them tilt their left ear over the sink or ground and flush from the left eye down. This reduces the chance of getting dirty water into the right eye.

By the way, you're probably going to make a mess if you're doing this indoors. Don't worry about it now. Focus on saving the person's vision. There will be time later for a careful mop-up.

If another person is nearby, have them call 911 when you flush the eye. If you're alone with the injured person, flush first, call second.

Cardiopulmonary Resuscitation (CPR)

Time for a commercial. CPR is a critical, life-saving technique that is administered to someone who:

- is unconscious,
- has stopped breathing,
- has no pulse.

In this situation, the person's heart has stopped. Their brain is still alive. The idea behind CPR is that you are manually pumping their heart and lungs in an effort to maintain the flow of oxygen, preserve brain function, and restart the heart. You only have about 3 to 5 minutes to resuscitate them and keep them alive until an ambulance arrives.

For this reason, I strongly urge you to take a CPR class. I assure you that there are countless ones available in your neighborhood, typically offered on a regular basis by the Red Cross or other organizations with certified instructors. Cost is nominal. These days it's possible to do "blended" instruction—a mix of online classes followed by hands-on training.

You will want hands-on training to effectively learn CPR. If you have not been trained, you may feel uncomfortable touching a stranger in such an intimate fashion. Or you may attempt to ape actions you have witnessed in television dramas, potentially making matters worse. The only reason they do it wrong on TV is that they're *pretending* to perform CPR on a living, breathing actor. They have to fake it, or they'd potentially hurt the person.

CPR is one of those things that even the professionals need to bone up on from time to time. Hospital staffers retrain each year because the reality of hospital work is that no one person performs CPR on patients when they arrive at the ER. The doctor might be checking their vitals, a resident might be doing chest compressions, a nurse stands ready to do timed breaths, and all the while, another person is prepping an IV.

All of us need to train, then retrain. Once you've been trained, you will know exactly what to do in the event someone has stopped breathing and you cannot find a pulse.

If you have not trained, you can still help an adult or teenaged victim by doing what's known as "hands-only CPR" or "hands-only compression." All you have to do is one simple thing: push on the adult's or teen's chest at 100 beats a minute. That may sound confusing, but I'm going to make it very simple for you. Do this: Go online and look up the song "Stayin' Alive," the 1977 disco hit by the Bee Gees.

No, I am not kidding. Go ahead. Watch any of the music videos you find online featuring that song. Let the song's magic sink into your brain. Pay attention to that infectious chorus (the part of the song that is repeated so many time that it sticks in your brain).

I'll wait...

That wasn't hard, was it? Here's a little secret: The rhythm of "Stayin' Alive" is precisely 100 to 120 beats a minute.

Now you know. In an emergency in which a person trained in CPR is not immediately available, you can compress a person's chest, up and down, to the tune of "Stayin' Alive," and you have a realistic shot at saving the person's life. If you are fearful of causing greater harm, please know that the Good Samaritan laws in the United States protect a person who attempts to save another's life. Only attempt this on a teen or adult. You will want to perform regular CPR—chest compressions and timed breaths—on infants and small children.

If this freaks you out, this might be your chance to get proper CPR training. And relegate the Bee Gees to Saturday night at your favorite club.

CPR on Infants and Small Children

Kids are smaller, so you need to use a gentler technique. First, check their airway for obstructions. Gently open their mouth, look inside, and pull out anything blocking their breathing. Always look first. Do not blindly probe; you can easily push something farther into their airway if you do that.

For infants: Place only the tips of your first two fingers on their breastbone, and pump down while thinking of the beat to "Stayin' Alive." One pump means their chest should only go down an inch. After 30 chest pumps, switch to timed breaths. Your adult mouth should cover both their little mouth AND nose at the same time, for a complete seal. Use a finger to gently elevate the child's chin. Blow into the child's mouth, stop, wait to see if the child breathes, do it again. After two breaths, switch back to chest compressions. Switch back and forth until help arrives.

For small children: Use both hands to press on the chest. Do 30 chest pumps before switching to timed breaths. Each pump can go down two inches. Pinch the child's nostrils closed while you cover their mouth with yours, and blow. Wait to see if they breathe. If not, do a second timed breath. After two breaths, then go back to 30 compressions, and alternate.

Choking

A person who is choking has food or some other object lodged in their throat. They will be unable to speak and may appear to be visibly distressed. Most of us have heard that the universal signal for choking is grasping one's throat with two hands. But not everyone has gotten the memo, and not everyone feels comfortable alarming others. If you suspect that someone is choking, speak up immediately: "Are you choking?" Get them to nod and indicate their predicament.

Two different steps are involved. Perform them in this order.

1. Ask them to stand up. Bend them at the waist slightly so they are facing the floor. Deliver five separate, open-hand blows to their back, right between the shoulder blades. The heel of your hand should be hitting the person's back in an upward manner. If that does not dislodge the object, proceed to the Heimlich maneuver.

2. Stand behind the person, and place a fist between the bottom of their ribcage and their belly button, thumb-side in. Grasp the pinkie-end of your fist with your other hand. Perform a quick upward thrust, almost as if you are trying to lift them off the ground. The purpose of the upward thrust is to force the remaining air in the person's lungs to expel the obstruction. Wait to see if the object is dislodged. If not, perform the upward thrust again. You may need to perform the maneuver several times to get the obstruction out. If the person passes out and you cannot feel a pulse, switch to CPR or hands-only CPR. Performing CPR will often dislodge the object in the person's throat.

Never perform the Heimlich on an unconscious person. Use CPR.

If, while attempting the Heimlich, you find that you cannot get your arms around a person due to their size, gently guide them to the ground and perform your upward thrust from above, in the same spot on their body as you would when they are standing: fist between ribcage and navel, aided by your other hand.

Cardiac Arrest/Heart Attack

Though laymen tend to lump these two together, they are absolutely not interchangeable.

CARDIAC ARREST is a malfunction in the heart's *electrical* function. The heart simply stops, or begins beating arrhythmically. The underlying causes are complex, and vary with every victim's medical background.

How it looks: The person may gasp, then stop breathing and pass out. You can safely rule out choking because a choking person will usually be grasping at their throat, and take much longer to pass out. A person with cardiac arrest is unconscious, and has no pulse and no breath. They need CPR.

First, make sure that you and the victim are in a safe location. You don't want to perform CPR in the middle of a road at night. Nor do you want to be on a balcony in a lightning storm, or anywhere near fallen power lines. If you must, move the person to a safe location. Act quickly and decisively.

You have two choices:

- If you're alone with the victim and have a mobile phone, dial 911 and call for help first. Then perform CPR. (Considering that most mobile phone calls can be placed on speakerphone mode, you could dial 911, switch to speakerphone mode, and speak to the 911 operator as you perform CPR. Only do this if it is not too complicated for you.)

- If other people are around, instruct *another* person to call 911 while you begin CPR. If you're in or near a commercial building, instruct bystanders to locate and bring an automated external defibrillator (AED) to your location.

Again, let me emphasize that you cannot hesitate, hem and haw, or wait for someone else to take charge. The victim may only have minutes to live. Begin CPR as soon as possible.

HEART ATTACKS are different. They are a loss of *circulatory* function. An artery in the heart has either become blocked, or "occluded," or is in the process of becoming more so. The classic signs are chest pains, feeling pressure in the body that begins in the chest and radiates down an arm (typically the left one), nausea, and vomiting. However, many victims—women in particular—have much more atypical signs. They may break out in cold sweats, feel intense jaw pain or back pain, or experience a sudden onset of debilitating fatigue.

Heart attacks are often not the dramatic, instant chest explosion we've all seen in the movies. Attack severity varies from person to person, and often depends on the specific artery that is experiencing impeded blood flow. The warning signs I just described might present themselves over hours, days, even weeks. Heart attacks can be slow, ticking time bombs.

In fact, when I was working in the ER in Colorado, our staff noticed to our horror that on nights of Super Bowl football games, heart attack victims routinely showed up at the hospital *after* the game. We didn't get many heart attack victims showing up while the game was in progress. This happened year after year, like clockwork—especially when the Denver Broncos were playing. Consider the implications: This means that people became excited during game play, began feeling uncomfortable, yet did not take action until the game's conclusion. I can only assume that a) they wanted to finish watching the game, or b) they didn't want to ruin the experience of their fellow sports fans!

You should never be so considerate! We have a saying in the medical profession when it comes to heart attacks: "TIME IS MUSCLE." That means the longer you delay treatment of a heart attack, the more heart muscle is damaged. Such damage is permanent. You don't regain that degree of heart function ever again. The muscle turns to scar tissue.

"TIME IS MUSCLE"

If you're experiencing any of these symptoms, speak up. If someone tells you they think they're having a heart attack, help them to remain calm. Exertion makes matters worse. Instruct someone to call 911. Heart attack victims are still conscious, so they do not need CPR. While you wait for medics to arrive, have the victim sit comfortably and loosen their clothing. Give them one aspirin—no more than one—and ask them to chew it. Yes, it will taste bad, but chewing will reduce the time it takes for the aspirin to be absorbed into the bloodstream. The faster it's absorbed, the faster it will stop blood clotting the victim's body. If you are in a location where waiting for an ambulance is impractical, arrange for a ride to an ER while the person takes the aspirin. It's better to have someone drive you and the victim so you can assist the victim during the drive, if necessary. If you attempt to help the victim while driving a vehicle yourself, the vehicle will quickly become a lethal weapon!

Drowning

Copious water in the lungs causes drowning. The person asphyxiates because they cannot get enough oxygen to sustain life. If you suspect someone is drowning, and they are still in the water, your best bet is to a) throw them a flotation device or rope to pull them out, or b) reach out to them with a pole, stick, or other object they can latch onto and be pulled from the water or moved closer to shore. Unless you are a trained lifeguard and equipped with life-saving equipment, never attempt to swim out to the person. A drowning person is a danger to others. Desperate and panicked, they can easily overwhelm an untrained person, shoving them under water in their own attempt to reach and stay at the surface. If you absolutely must dive in to rescue someone, it is best to approach them from behind, not head-on, so they cannot see you coming and cause additional problems.

Once you have the person safely on the ground, assess if they are still conscious. If so, rest them on their side. They will naturally begin coughing up the water in their lungs. Assist them so they don't fall onto their backs. You want the water and possibly vomit to land right in front of them, not back into their mouths.

If they become unconscious, check for a pulse and start CPR if they have stopped breathing.

Seizures

Seizures are caused by abnormal electrical activity in the brain, and can be frightening events for sufferers as well as family members or passersby who feel compelled to help in some way. If a person has a history of epilepsy, they have a chronic neurological impairment in which a cluster of nerve cells in their brain will fire improperly at the same time, triggering a seizure. In diabetics, the abnormal electrical brain activity is triggered by inconsistent sugar levels in their body.

Outwardly, these types of seizures will manifest themselves in one of two ways. Diabetics and epileptics typically exhibit what doctors call a *grand mal* seizure, characterized by a loss of limb control, flopping around, a loss of bladder control, etc. This type of seizure has been popularized so much in movies that even people who have never witnessed one will tend to associate this behavior with seizures, or erroneously, strokes. The *petit mal* seizure, the second type, is less dramatic but no less serious. The victim will cease all speech or action, and simply gaze off unresponsively into the distance.

Both types of seizures can be dangerous, especially if the person is out of doors, where sudden loss of action or flopping can pose a threat to their safety. To help a seizure victim, call 911 at once. Protect the person's head from further harm. Don't restrain arms and legs; just keep watch on them so they don't cause the person to injure themselves further. If you can guide their head onto your lap, a soft cushion, or a rolled-up jacket, that will suffice.

Don't insert anything into the person's mouth. This is bad advice, often suggested to keep the person from biting their own tongue. Tongue-biting doesn't happen often, because the victim's body is far too rigid during a seizure to bite. If they do bite their tongue, it will happen within the first few seconds of the seizure and you'll be unable to do anything about it. If the

STROKE, SEIZURE, HEART ATTACK, CARDIAC ARREST: Which Is Which?

Is the person . . .

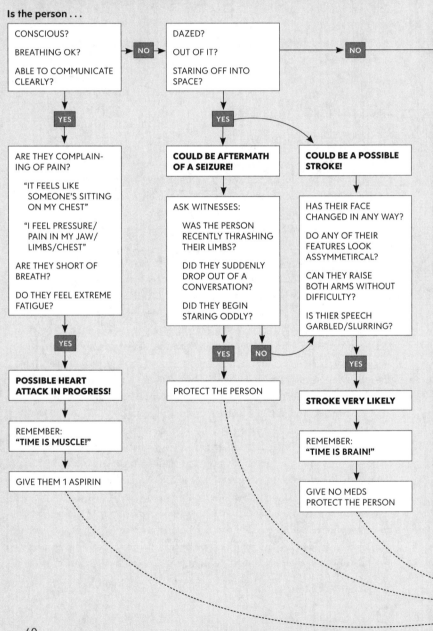

CONSCIOUS?

BREATHING OK?

ABLE TO COMMUNICATE CLEARLY?

NO

DAZED?

OUT OF IT?

STARING OFF INTO SPACE?

NO

YES

YES

ARE THEY COMPLAIN-ING OF PAIN?

"IT FEELS LIKE SOMEONE'S SITTING ON MY CHEST"

"I FEEL PRESSURE/ PAIN IN MY JAW/ LIMBS/CHEST"

ARE THEY SHORT OF BREATH?

DO THEY FEEL EXTREME FATIGUE?

COULD BE AFTERMATH OF A SEIZURE!

COULD BE A POSSIBLE STROKE!

ASK WITNESSES:

WAS THE PERSON RECENTLY THRASHING THEIR LIMBS?

DID THEY SUDDENLY DROP OUT OF A CONVERSATION?

DID THEY BEGIN STARING ODDLY?

HAS THEIR FACE CHANGED IN ANY WAY?

DO ANY OF THEIR FEATURES LOOK ASSYMMETIRCAL?

CAN THEY RAISE BOTH ARMS WITHOUT DIFFICULTY?

IS THIER SPEECH GARBLED/SLURRING?

YES

YES

NO

YES

POSSIBLE HEART ATTACK IN PROGRESS!

PROTECT THE PERSON

STROKE VERY LIKELY

REMEMBER: **"TIME IS MUSCLE!"**

REMEMBER: **"TIME IS BRAIN!"**

GIVE THEM 1 ASPIRIN

GIVE NO MEDS PROTECT THE PERSON

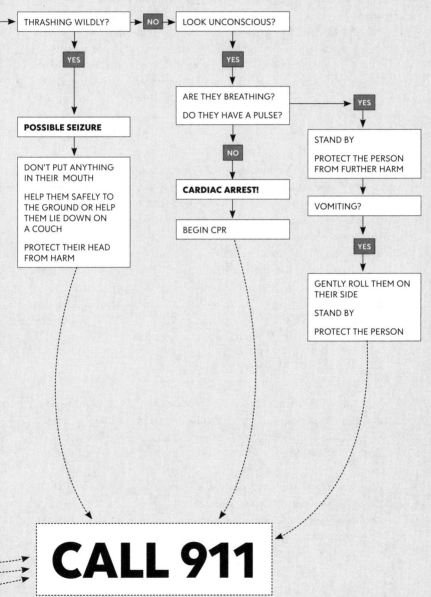

THRASHING WILDLY? → NO → LOOK UNCONSCIOUS?

YES

POSSIBLE SEIZURE

DON'T PUT ANYTHING
IN THEIR MOUTH

HELP THEM SAFELY TO
THE GROUND OR HELP
THEM LIE DOWN ON
A COUCH

PROTECT THEIR HEAD
FROM HARM

YES

ARE THEY BREATHING?
DO THEY HAVE A PULSE? → YES

NO

CARDIAC ARREST!

BEGIN CPR

STAND BY

PROTECT THE PERSON
FROM FURTHER HARM

VOMITING?

YES

GENTLY ROLL THEM ON
THEIR SIDE

STAND BY

PROTECT THE PERSON

CALL 911

person is bleeding from their mouth, tilt their head so the blood drains to the ground and not back into their mouth.

Be patient. Most episodes last 30 seconds to 2 minutes. When the seizure stops, the person may experience about 10 minutes of confusion, or possibly longer. They may not speak, and simply be trying to make sense of what happened.

In both situations, the person ought to be seen immediately by a doctor. With diabetes, there is a possibility that the seizure will result in a coma. And a person with epilepsy may have an acute episode that lasts longer than 30 minutes, called *status epilepticus,* that needs to be treated with medication.

Stroke

In the fall of 2020, an 85-year-old American politician suffered what appeared to be a stroke while doing a live interview on his social media channel. In the middle of a conversation with an interviewer on the topic of economics, the former congressman's face appeared to change instantly. The left side of his mouth drooped, and he began slurring his words. He displayed, in other words, two of the classic signs of a stroke.

You'll recall that in heart attacks, doctors say, "TIME IS MUSCLE," meaning that delaying medical attention risks the damage of more heart tissue.

With strokes, doctors say, "TIME IS BRAIN." Time wasted means more brain damage. So you really can't waste time asking people if they are okay, if they'd like a drink of water, if they'd like to rest or lie down.

"TIME IS BRAIN."

No to all of that! Strokes are all about blood circulation. For some reason, the person has experienced a clot in a blood vessel in the brain, or one of those vessels has started to bleed. A stroke is significantly different from a seizure. Doctors use the acronym FAST to discuss treatment of dealing with a stroke:

Face: Has the person's face changed? There may be some drooping of the features, or an asymmetrical appearance.

Arms: Ask them to raise both arms. Is one drooping or weaker than the other? (A stroke victim might not be able to lift the arm that was impacted by the stroke.)

Speech: Is the person's speech garbled or slurring?

Time: Time is critical. If you can answer yes to any of the above points, you must call 911 immediately. Without immediate medication, the victim risks losing more of their brain function. The medication must be given as quickly after the stroke as possible.

Do not give the person aspirin. That could make the brain-bleed worse. Assess the situation, think clearly, and get help.

THE FINE ART OF ADVENTURE PACKING

O ver the years, I've developed a sixth sense about how to pack for my various trips. If I'm going to Paris for work, I know that medical attention for a serious emergency is only a phone call away. In addition to my passport, I'll toss a small first-aid kit and some over-the-counter painkillers in my suitcase, just in case, so I don't have to go scrounging through Parisian pharmacies for Le Tylenol and Le Band-Aids when I should be ordering a bottle of Sancerre and some stinky cheese at a local bistro.

On the other hand, if NBC is sending me to the Amazon jungle, I know I should go full Adventure Doc, and bring along my most complete medical kit, not to mention some intravenous fluids in case I need to treat myself or a crew member when we are far from a major city.

The great outdoors—even the state park close to home—can turn into that jungle if you're not prepared. Accomplished outdoors people invest in quality gear and may have participated in training courses before they embark on long wilderness adventures. If they have been hours or days from civilization before, chances are they have absorbed a good deal of knowledge about good safety practices.

A good rule of thumb: The longer or farther you will be from civilization, the more supplies and the more skills you should have to be able to take that trip.

The type of people who turn up in emergency rooms on the weekends are those who decided on a lark to do a short hike, rent a kayak for a short paddle, or otherwise engage in some activity outside their normal range of experience or skill level.

"It's just a short trip," they tell themselves. "What could possibly go wrong?"

I'm here to tell you that things can and do go wrong, even in the shortest timeframe. The classic is the day hike that for some unforeseen reason extends until dusk or well after dark, because one or more people misjudged the level of exertion or the amount of sunlight left in the day.

Bottom line: Learn to pack appropriately.

What follows is my basic checklist for stuff to carry in your backpack on what I hope does *not* turn into a never-ending hike.

Day Hike Pack List

- ☐ Blister or moleskin bandages (i.e., Band-Aid Hydro Seal, Compeed, etc.)

- ☐ Cold-weather clothing (fleece jacket, extra layer or shirt, etc.)

- ☐ Duct tape

- ☐ Fire starter tool

- ☐ Food and water for each person

- ☐ Headlamp

- ☐ Insect repellent

- ☐ Knife or multitool

- ☐ Lightweight emergency blanket

- ☐ Navigation tools (compass, map, GPS device)

- ☐ Small first-aid kit

- ☐ Sunscreen

- ☐ Tweezers (for splinters or tick removal)

- ☐ Waterproof matches or lighter

- ☐ Waterproof notebook and pen (leave a note at the trailhead indicating your time of departure and expected return)

- ☐ Whistle (three loud blasts is the universal code for HELP)

Going Full MacGyver

MacGyver was the hero of an '80s TV show who always knew how to repurpose objects in his surroundings to save lives or get out of a hair-raising situation.

It's pretty obvious that you will never be able to bring or pack every single thing you need in an emergency. But if you calm down and think logically, you'll find that many objects can be used to perform a function.

- Duct tape can close a wound and keep it shut until you reach a hospital.

- Hand sanitizer makes an excellent fire starter because of its alcohol content.

- A tiny strip of aluminum foil held to each end of a battery will generate a spark as strong as any piece of campsite flint. But you must be ready with a flammable piece of kindling to catch that spark.

- No kindling? The cotton fiber in tampons is the perfect substitute. (They stop nosebleeds too.)

- Condoms can serve as makeshift water bottles, slingshots, or finger and wrist tourniquets. If you have enough of them, tie them together to bind someone's broken hand, finger, or arm to a splint.

If You Get Lost in the Woods

Every year people get lost and die in the wilderness. Some even become hopelessly lost while in their *vehicles* because they relied more on their GPS systems than on actual maps and common sense. A hand-held device is often no substitute for a solid grasp of wayfinding. Silicon brains have a bad habit of losing their signals, sending travelers down roads that are unfit for ordinary vehicles, or failing to keep abreast of all weather conditions.

If you're going on foot, commit to sticking to the trail. If the trail becomes unclear or ambiguous, it's time to head back. Leaving a note at the trailhead before you embark on your hike is a must. But so is letting a trusted person know where you're going and when you'll be back. A note should spell out the time and date you left, the number of people in your party, their names, emergency contact information, your intended route, and return time. This is the only way rangers or other authorities have of knowing someone might need help. (Always remove the note upon return.)

You should always have a rough idea of the compass direction you need to follow in order to return to your campsite, civilization, or the place where you left your vehicle. With this in mind, you can use your compass and paper maps to find your way. Or use the wristwatch method (see pages 69–70) to head in the right direction.

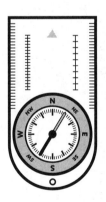

If these efforts prove unhelpful, it's smarter to *stop* moving. Search and rescue teams use a grid system to locate lost hikers. They sweep through one box of a map grid, cross it off their list, and move on to the next box. If you keep moving, you defeat this system. A common saying in this world is: "If you're lost, hug a tree." That means stay put, seek the best shelter you can, and hunker down until someone finds you. (You don't really have to hug a tree, but you get the idea.)

In summer 2020, mariners who became stranded on a small deserted atoll in the Pacific were found by rescuers who saw the SOS distress signal that they spelled out in the sand with beach debris. Sure, they could have spelled out the word HELP, but SOS was devised in the early 20th century specifically because it didn't require people in distress to be able to speak the language of all possible rescuers in the vicinity. SOS is an internationally known symbol for distress. You can spell the letters out on the ground. You can beat an audible signal on rocks. You can use a mirror in daytime to flash the code from a prominent spot in the wilderness. And by night you can use a flashlight to send a distress call over long distances. The code is simply three short flashes, three long, three short (* * * — — — * * *). The complete Morse code alphabet is shown here, but don't get hung up on trying to send complex messages. SOS is the quickest and best way to call for help.

International Morse Code

Letter	Code	Letter	Code	Signal	Code
A	• ▬	N	▬ •	Starting Signal	▬ • ▬ • ▬
B	▬ • • •	O	▬ ▬ ▬	End of work	• • • ▬ • ▬
C	▬ • ▬ •	P	• ▬ ▬ •	Error	• • • • • • • •
D	▬ • •	Q	▬ ▬ • ▬	1	• ▬ ▬ ▬ ▬
E	•	R	• ▬ •	2	• • ▬ ▬ ▬
F	• • ▬ •	S	• • •	3	• • • ▬ ▬
G	▬ ▬ •	T	▬	4	• • • • ▬
H	• • • •	U	• • ▬	5	• • • • •
I	• •	V	• • • ▬	6	▬ • • • •
J	• ▬ ▬ ▬	W	• ▬ ▬	7	▬ ▬ • • •
K	▬ • ▬	X	▬ • • ▬	8	▬ ▬ ▬ • •
L	• ▬ • •	Y	▬ • ▬ ▬	9	▬ ▬ ▬ ▬ •
M	▬ ▬	Z	▬ ▬ • •	0	▬ ▬ ▬ ▬ ▬

Punctuation	Code
.	• ▬ • ▬ • ▬
,	▬ ▬ • • ▬ ▬
?	• • ▬ ▬ • •
'	• ▬ ▬ ▬ ▬ •
/	▬ • • ▬ •
:	▬ ▬ ▬ • • •
;	▬ • ▬ • ▬ •
+	• ▬ • ▬ •
-	▬ • • • • ▬
=	▬ • • • ▬

THE RULE OF THREE

You have probably heard that if you're lost in the wilderness, your first concern should be to seek shelter, water, and food—in that order. This is absolutely correct. Most of us are so accustomed to being able to eat and drink whatever we want, whenever we want, that we incorrectly assume that food and water are absolutely critical to survival. Nope. You can go quite a while without potable (i.e., clean) water, and even longer without food.

Humans are designed to store fat for that very reason. Our ancient ancestors were built to travel long distances in search of fresh water and game to hunt. But every day, they were obliged to seek shade in hot environments, build fires, or huddle in animal skins in caves to stay shielded from the wind, rain, and cold.

Now, don't get me wrong. You will feel miserable without water very quickly, but you'll survive—at least for a little while. Ditto for food.

To express these truths about human physiology, wilderness and survival experts often refer to a rubric called the Rule of Three. The Rule summarizes in a nutshell your likelihood of surviving extreme situations.

The Rule of Three goes like this:

A human can live . . .

3 minutes without oxygen (or in cold water or bleeding heavily)

3 hours without shelter (in extreme cold or hot weather)

3 days without water

3 weeks without food

Notice that the further down the list you go, the more *time* you have to survive, but each step down the list assumes that you have easy access to the things that come before it. If you don't have shelter, you're not going to survive to find water. If you don't have shelter or water, you're not going to live long enough to rustle up some grub.

I want to call your attention to the very first rule. It's the one I came up against all the time when working in a hospital emergency room. If a person goes into cardiac arrest—their heart stops cold—they really only have 3 minutes to get their blood flowing again. Five minutes, tops. Beyond that, they're looking at severe brain damage, even if doctors, medics, or bystanders are able to get the person's heart started again.

This underscores the importance of learning CPR and being prepared to administer it at a moment's notice. Precious minutes that could save a person's life are often wasted while bystanders wait around for an ambulance to arrive. Now you know: When emergency strikes, it's better to act than to wait.

None of us can beat the Rule of Three.

HOW TO EVACUATE

The first and only time my family had to evacuate our home in Colorado, it was because the wildfires raging across the landscape had defied all predictions of the scientists, meteorologists, and firemen. Nature did what the experts thought it wouldn't or couldn't.

Color me shocked. Nature is unpredictable. Who knew?

When the authorities announced that they wanted us out, I was still at work. As I finished up my shift, I got a call from a vet friend. His medical facility was closing up and sending their employees home, but he also boarded animals for families on vacation. He needed help. Would I come help him load all those dogs and cats into a vehicle for safe evacuation?

I was fine with that. But before I headed over to help him, I just checked with my wife, to make sure she was preparing herself and our daughter to leave. Lynda said everything was in good shape. She had told our daughter, who was then on college break, to pack carefully for a trip out of town. (Our son was safely at school in another region of the country.)

By the time I got back to the house, I couldn't help but notice the differences in the way my wife and daughter had packed. Lynda had packed like the great mom she was: making sure we had food, water, clothing, a solid first-aid kit, blankets, pillows, and tools, just in case we needed them on the road.

My daughter had packed the car with sentimental items she kept in her room. Stuffed animals, letters from families and friends, and photo albums.

It taught me a lesson. There are always different ways to evacuate. Different philosophies. Some people pack as if they're merely going on a 3-day trip. Others pack as if they're never coming back. In our case, the wildfires were controlled in a matter of 5 days. But there was a germ of wisdom to my daughter's approach. It was entirely possible that we'd never see our home again. In that case, she was smart to bring along her most precious items.

Some tips for evacuating...

- It really pays to get out as soon as the authorities in your area suggest or mandate evacuation. In my experience, public officials hate raising false alarms because it simply makes them look like the Boy Who Cried Wolf. The next time a serious evacuation is necessary, people will ignore their orders. If they say it, they really mean it. Take them at their word. We all know stories of people who waited too long to leave before an impending weather event or natural disaster, or who refused to leave altogether, and end up stranded or worse. The longer you wait to get out of Dodge, the more traffic you encounter, the fewer choices you have for fuel, lodging, food, etc. In our case, we could not find a hotel room, and were forced to bunk with a cousin in Denver. We will always be grateful that Luis and Ida were willing to help us out of that predicament!

- Think about people first. For example, if you have elderly loved ones in the area or ones with special needs, you'll want to deal with their safety first. Decide if they are evacuating with you, or if you are entrusting their care to others.

- Next, decide how/if your pets are traveling with you. If you have livestock, ideally you have made contingency plans well in advance; now's the time to activate them.

- The best packing list is one you have literally written down months ago. This way, you are unlikely to become rattled and forget critical items. (See my working list on pages 55–60.)

- It is always wise to take important documents such as insurance policies, deeds, etc.

- Before you leave home, shut off your water, natural gas lines, and any other utilities that might pose a danger to others if the worst comes to pass.

YOUR BUG-OUT "BAG"

T he notion of the bug-out bag is lifted straight out of the world of military preparedness. The concept is simple: If you have to flee your home and hit the road because of a disaster, you ought to be able to live on your own before help arrives. In most developed nations, the cavalry will get on the scene in 1 to 3 days. So: If you lost power and water and other essentials for up to 72 hours, what would you need to get by? If you had to evacuate, what would you take?

As I've hinted elsewhere, it's really easy to *ask* these questions, but tough to drill down on perfect answers. Every situation is different, as are people's needs and skill levels. What you bring with you in case of a wildfire evacuation is different from what you'd take along in case of a predicted hurricane, flood, or blizzard. And everything depends on your mode of transportation. If you're packing everything into the SUV and hitting the highway, what you can bring along is very different from what a person or family can carry on foot. Be careful. Access to a vehicle encourages people to bring everything but the kitchen sink. Before you pull out of the driveway, consider how you can re-prioritize your bug-out supplies and move nimbly on foot if your car breaks down or must be abandoned.

These are hard things to consider, and I don't pretend to have answers that will satisfy and help everyone who picks up this book. With the list that follows, my goal is simply to get you to think about what you could possibly bring, and what might be helpful in a pinch. I encourage you to adapt the list to your needs. That's the only way it will be useful to you.

On one hand, what I've compiled looks like a packing list for an awesome camping vacation. A lot of the gear—like a first-aid kit or sewing kit—should be in your home right now, because it could actually be useful any day of the week (see page 20). And most of the gear will come in handy if you simply have to hunker down in your house for days without power, water, or heat while waiting for those important services to be restored. Between a local supermarket, a good outdoor store (i.e., REI, Bass Pro Shops, Cabela's, Dick's Sporting Goods), and an online retailer (i.e., Amazon), you should be able to assemble much of this equipment.

You will notice some redundancy in this list as well. That's a *good* thing. You always want to have more than one way to perform the same critical task. If a can opener breaks or is lost, it helps to know you have a backup one on your Swiss Army knife or multitool. More than one flashlight is never bad—as long as they work.

A final note: I put the word *bag* in quotes because clearly not all of this stuff will fit in a single carrying case. To save time in an emergency, you might want to pre-pack some of the non-perishable and not-likely-to-be-used-today items in a duffel or backpack. Store it in the garage or a closet so it can be easily grabbed when it's time to get out of Dodge.

Your Prospective Bug-Out "Bag" Checklist

☐ **Backpack or waterproof bag**
You need something to carry it all in, don't you?

☐ **Dehydrated meals (i.e., brands such as Mountain House, Readywise, Backpacker's Pantry)**
Easily found at outdoor stores and many warehouse clubs, these "just-add-hot-water" meals can last 2 to 5 years on the low end, 25 years on the high end! You'll want enough for 3 to 5 days for each person.

☐ **High-calorie or high-protein meal bars**
Shorter shelf life, but you won't need to stop and boil water.

☐ **Trash bags**

☐ **Water**
Humans need 1 gallon per person a day.

☐ **Can opener**

☐ **Camp stove, pots and pans, and cleaning supplies**
Classic camp stoves and grills (i.e., Coleman) run on small, portable, refillable butane or propane tanks; newer types (i.e., Solo Stove) burn twigs and pine needles, so you're not dependent on fuel access. All are available either at an outdoor store, or via online retailers. Aluminum foil can be handy for certain meal preparations. Just saying.

☐ **Eating utensils**

☐ **Water bottle (i.e., Nalgene or stainless steel)**

☐ **Water filter or filtering system, iodine solution, water purifying tablets**
See how and when to use these on page 75.

☐ **Hand-powered radio**
AM, FM, and NOAA weather are musts. Some brands have built-in flashlights and USB chargers. Some run off solar power; some are hand-cranked.

☐ **Work gloves**

☐ **Knife**
Look for a "combo" knife that has a partially serrated blade.

☐ **Paracord rope**
Easy to cut, tie, or braid for multiple uses.

☐ **Multitool or Swiss Army knife**
Some are equipped with handsaws and whistles, eliminating the need for the next two items.

☐ **Whistle**

☐ **Handsaw**

☐ **Combo folding shovel/pickaxe**
Use to handle hot coals, dig holes, dislodge rocks or debris, etc.

☐ **Combo axe/hammer**
Chop wood, pound in tent stakes.

☐ **Duct tape and Ziploc bags**
Both are insanely useful. Choose an assortment of bag sizes—pint, quart, or gallon. Keep in mind that freezer and storage bags are more durable than sandwich-style bags.

☐ **Compass**
The world's most essential navigational tool.

☐ **Wristwatch**
In case your mobile phone or vehicle clock are unavailable. Use to track heart rates or respiratory rates, or record how long ticks have been on a person's skin. Choose a tough field or dive watch with good water resistance. Analog watches can be used as a compass in a pinch. (For more on the ideal disaster watch, see page 70.)

☐ **Maps**
Of the local area and larger geographical region.

☐ **Waterproof notebook, pens, pencils**

☐ **Weather gauge system (i.e., Tempest, AcuRite, etc.)**

☐ **Important personal documents**
Driver's licenses, medical prescriptions or records, critical financial and insurance papers, passport, birth certificates, etc.

☐ **Books, decks of cards, games, puzzles**
Anything that will help your family pass the time.

☐ **Laptop computer(s), charger, and backup drives**

☐ **Mobile phone(s)**
Always pack a charger or two. Remember that when reception is spotty, a text is often the smartest way to get a message to someone in a hurry, because texts don't require as much of a signal to transmit as calls do. And texts don't eat up as much precious battery time.

☐ **Mesh network device (i.e., goTenna)**
Allows you to use your mobile phones to send texts and
your GPS locations over long distances with others in your
party, even if cell towers are not available. The underlying
technology is similar to walkie-talkies.

☐ **Small amount of cash, in small bills**
In case bank ATMs are not available.

☐ **First-aid kit**

☐ **Personal, wearable ID band**
One per person, listing emergency contacts and/or medical
information for those with life-threatening conditions—not to
mention everyone else. Identification becomes critical if your
family becomes separated, an injury prevents communication,
and/or personal documents are lost.

☐ **Prescription meds and over-the-counter drugs**

☐ **Stormproof matches, liquid fuel lighter**

☐ **Fire starter tool**
In case you lose, damage, or finish your matches. See how to
use one on page 64.

☐ **Disposable hand warmers**

☐ **Sewing kit**
Safety pins can be used as fishhooks in a pinch. Needles can
mend clothing or stitch human skin.

☐ **Fishhooks, line, yo-yo reel**
A "yo-yo" fishing reel is simply a plastic reel around which you
wrap your line. It's an inexpensive item that will not take as
much room in your bag as a fishing rod. You can easily locate
videos online demonstrating their use.

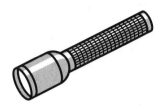

☐ **Superglue adhesive**

☐ **Toiletry kit**
Razors, body wipes, tampons, condoms, toilet paper, etc. A portable mirror can be used to signal across long distances in an emergency.

☐ **Soap**
Gentle castile soap (i.e., Dr. Bronner's) will wash nearly everything—your body, clothes, or cookware.

☐ **Hand sanitizer**

☐ **Antibacterial or sanitizing wipes**

☐ **Sunscreen**

☐ **Insect repellent**

☐ **Bear spray/pepper spray**
No guarantee of success, on either human or animal foes. Be sure you have trained in the use of these products, and know how to discharge them in a safe manner.

☐ **Flashlight**
Bring spare batteries. Using the flashlight on your crank radio is hard work.

☐ **Headlamp**
To keep your hands free if you are forced to walk in darkness, whether in a building, city street, or outdoor trail. Battery-powered, solar, or chargeable via USB port.

☐ **Portable jumper cables**
Modern ones, equipped with a battery, can also quickly charge mobile phones.

☐ **Solar panel and charging battery**
To charge your phone when electricity is not available.

☐ **Goggles and face masks**
A set for each person, to keep smoke, dust, and toxic airborne particles out of your eyes, nose, mouth, and lungs.

☐ **Clothing**
Hat, gloves, beanie, socks, fleece, rain poncho, performance layer shirt, underwear, long johns.
    You'll want more layers than you think, even in warm seasons. "Performance" outdoor clothing is durable, hand-washable, and dries quickly. Wool will keep you warm even when wet. Cotton—however breathable—will not.

☐ **Tent**

☐ **Sleeping bag/pad**
One per person.

☐ **Tarpaulin**
For under your tent.

☐ **Emergency blanket**

☐ **Mountain bike(s)**
They can go where your car can't or won't, if fuel is no longer available.

☐ **Inflatable pack raft**
In case of floods, but you'll need a hand or battery-powered pump to inflate. If you go this route, make sure you have one large enough for your party, and oars to steer it.

SHOPPING FOR YOUR BUG-OUT BAG

Don't go overboard. Buy only what makes sense for your family and circumstances. Choose items that you know you will use in the future, not just in emergencies. If you're an outdoorsy family, you may have many of these items already. Optional "gonzo" items are marked with an asterisk (*). Some of these are high-ticket items, so choose appropriately.

Snap a photo of this checklist so you can refer to it on your mobile device as you shop. Or scan it and print a hard copy to take with you. I assure you that any unusual items that you cannot find in stores are available via online retailers.

Supermarket
- [ ] Batteries for all your devices
- [ ] Bottled water (in gallons or smaller)
- [ ] Can opener
- [ ] Food (when needed)
- [ ] Hand sanitizer
- [ ] Insect repellent
- [ ] Liquid dish soap
- [ ] Liquid fuel lighter
- [ ] Paper plates/plastic utensils (or washable camp plates/utensils at outdoor store)
- [ ] Paper towels
- [ ] Pens/pencils
- [ ] Sanitizing wipes
- [ ] Scouring pads

- [ ] Sewing kit
- [ ] Snack bars (high-calorie or high-protein)
- [ ] Soap
- [ ] Sunscreen
- [ ] Superglue
- [ ] Toiletries (razors, wipes, tampons, condoms, toilet paper)
- [ ] Trash bags
- [ ] Ziploc bags (freezer/storage, pint, quart, or gallon)

Drugstore
- [ ] 30-day supply of your family's prescriptions
- [ ] First-aid kit
- [ ] Over-the-counter drugs

ATM
- [ ] Cash for emergencies

Hardware Store
- [ ] Combo axe/hammer
- [ ] Folding shovel/pickaxe
- [ ] Handsaw
- [ ] Lighters
- [ ] Matches
- [ ] Tarpaulin
- [ ] Work gloves

Outdoor Store or Online Retailer
- [ ] Backpack or waterproof bag(s)
- [ ] Bear spray
- [ ] Camp stove and fuel
- [ ] Compass

- [ ] Dehydrated meals (3 to 5 days per person)
- [ ] Emergency blanket
- [ ] Face masks
- [ ] Fire starter tool
- [ ] Fishing equipment (hooks, line, yo-yo reel)
- [ ] Flashlight
- [ ] Fleece jackets
- [ ] Goggles
- [ ] Hand warmers
- [ ] Hand-powered radio
- [ ] Hats/beanies
- [ ] Headlamp
- [ ] Inflatable pack raft*
- [ ] Iodine solution
- [ ] Knife (folding, with both serrated/straight edge)
- [ ] Long johns
- [ ] Long pants/shorts
- [ ] Maps (local and regional)
- [ ] Mountain bike*
- [ ] Multitool or Swiss Army knife
- [ ] Paracord rope
- [ ] Pepper spray
- [ ] Performance layer shirt
- [ ] Plates and utensils (if you don't want disposables)
- [ ] Pots and pans
- [ ] Rain poncho
- [ ] Sleeping bag/pad
- [ ] Socks
- [ ] Stormproof matches
- [ ] Tent
- [ ] Underwear
- [ ] Warm gloves
- [ ] Water bottle(s)

- [ ] Water filtering system
- [ ] Water purifying tablets
- [ ] Waterproof notebook
- [ ] Whistle
- [ ] Wristwatch

Personal

- [ ] Birth certificates
- [ ] Computer and hard drives
- [ ] Driver's licenses
- [ ] Financial documents
- [ ] Insurance documents
- [ ] Medical prescription documents
- [ ] Medical records
- [ ] Mobile phones
- [ ] Passports
- [ ] Phone chargers
- [ ] Recreation: books, cards, games, puzzles

Go Gonzo

- [ ] Battery-charger jumper cables* (https://no.co)
- [ ] Mesh network device* (https://gotennamesh.com)
- [ ] Personal wearable ID band* (https://www.roadid.com)
- [ ] Solar panel and charging battery* (https://www.goalzero.com)
- [ ] Weather gauge system*
- [ ] (https://weatherflow.com/tempest-weather-system/)
- [ ] (https://www.acurite.com/shop-all/weather-instruments.html)

HOW TO USE
A FIRE STARTER

F ire is humankind's oldest friend. It keeps us warm, keeps predators at
bay, lights the way in darkness, and cooks our food. If you lose power or
you are forced to rough it outdoors, you will always be better off if you
have an abundant source of firewood and know how to start a fire. In a pinch,
a fire just makes you *feel* okay, like you've managed to retain some semblance
of human civilization.

Squatting over your campsite, rubbing two sticks together, is not the
way to go. That takes far too much time. For a fun, brief family camping trip,
you're probably bringing along a box of matches tucked into a plastic bag, or
a liquid fuel lighter to light your campground grill. That's not good enough
for disaster planning.

Why? On short camping trips, you have probably noticed how quickly
everything you've lugged along with you gets wet, or at least damp. Luckily,
your clothing, your tent, your sleeping bag are still absolutely serviceable
when wet. Everyday paper or wooden matches, not so much. If they get wet,
they're useless. Experienced campers pack stormproof matches that are so
rugged that they can stay lit in wind and rain. (Some are so good they stay lit
underwater!) For backup, serious campers bring multiple lighters. As a last
resort, they pack some fire starter tools.

People refer to these as "flints," but they are really made of metal. One
type is magnesium bars. The other type, known as "ferro" rods, is made of an
alloy called ferrocerium (which also has a little of that magic element, magnesium, in it). Metal fire starters are very durable, and will still work even when
they have been wet. In general, you get what you pay for. Buy the best quality rods and bars you can find. Cheaper rods, for example, may not contain
enough magnesium to do the job.

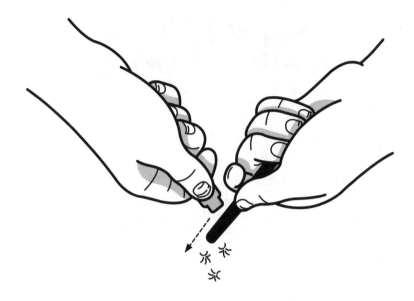

How to use fire starter tools:

- Gather your kindling. You can use any combustible material as long as it's dry. Man-made objects like toilet paper, paper towels, notebook paper, paper plates, and tampon cotton can all work. A lot of backpacking and camping equipment employs a nylon-covered rope called paracord. Some brands have fluffy white inner strands that are combustible, thus making great tinder. Objects found in nature like twigs, small sticks, tree bark, and cattails are the original kindling.

- If it has recently rained, don't despair! Dead plant matter is superficially wet, but still quite dry inside. You can, for example, push away a damp layer of pine needles or leaves in a forest to find a drier layer below it. You can break dead, drip-dried branches directly off a tree. You can use a knife to scrape tree bark, or scratch up the inner portion of a tree stump, to access its dry dead matter. Arrange all of this kindling in a tiny circle.

- Assemble some larger pieces of wood that you will use to feed the fire once it gets started. Keep them handy.

- Use the scraper that came with your ferro rod to remove the rod's black protective coating, exposing the silvery alloy beneath it. Next, digging a little harder with the serrated edge, scrape a few filings of the ferro off the rod. Let them fall into the center of your kindling. (Scrape some filings if you're using a raw magnesium bar.)

- Next, with a quick, downward motion, flick the smooth edge of the scraper down the length of the ferro rod. (Or use the designated striking surface of your magnesium bar.) You will get sparks! Aim them at the pile of tinder and metal filings. You might need to experiment to get the right angle. Often a short stroke will produce more focused sparks than a long stroke.

- Magnesium burns bright and hot! Those flames will almost certainly ignite even the dampest kindling. Just make sure your hair, fingers, and clothing are not in the zone of ignition. Once the sparks turn to flame, blow to stoke the fire and add more fuel as necessary.

HOW TO USE YOUR WRISTWATCH TO FIND YOUR WAY

You'd be surprised how quickly you can lose your way on an afternoon hike, especially if you wander away from the trail to capture a lovely scene with your camera, or to investigate a water feature. It's equally bad when disaster strikes, and you are stranded in a location far from home. You may have a compass app loaded onto your phone, but if you've drained the battery, that handy piece of technology is next to useless.

Wayfinding is a critical skill. Everyone needs to be able to find their way from here to there, from north to south, from the idyllic picnic site in the woods to the spot where you left your car. I always recommend that people buy a dedicated, high-quality compass that they can tuck into a backpack when they are on the go.

These days, compasses may feel like antiquated technology, but they can be lifesavers. Please don't skimp on these essential devices. Manufacturers like to build small compasses into everything. Your fleece jacket may have one built into its zipper. I once bought a fire starter tool that had a combo compass and whistle built into it. They're handy, but please do not entrust your family's survival to such flimsy tech. It always makes sense to buy a high-quality, dedicated compass at a professional outdoor store, and spend a few minutes with the retailer learning how to use it. Most of the salespeople in those places love what they do, and can easily assist you.

If your afternoon hike goes horribly wrong, you have three ways of finding the right direction. I'm listing them here in order of each technique's relative accuracy.

1. You can use your compass.

2. You can use the military-approved stick-and-shadow method. Here's how that works:

   a. We all know the sun rises in the east and sets in the west. If it is morning and sunny, and you stick a tall, straight stick into the ground, and watch where it throws its shadow, the shadow will be cast in a westerly direction. Why? Because the east-rising sun is headed west. Mark that position with a stone or a second stick.

   b. But watch! If you wait 15 or 20 minutes and check the shadow again, it will have moved away from your stone or second stick in a more easterly direction than it was before. Why? Everything's relative. The more west the sun goes, the more eastward it is throwing its shadow. Get yourself a third stone or long stick and mark that spot.

   c. You can do this all day, marking points along the eastward-moving line. But you have places to go. From here, imagine a perfectly

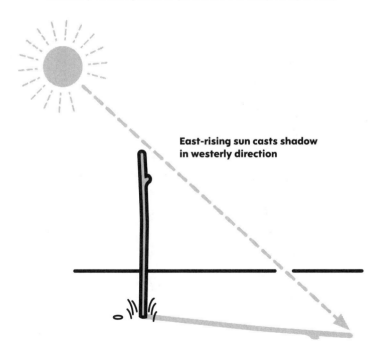

East-rising sun casts shadow in westerly direction

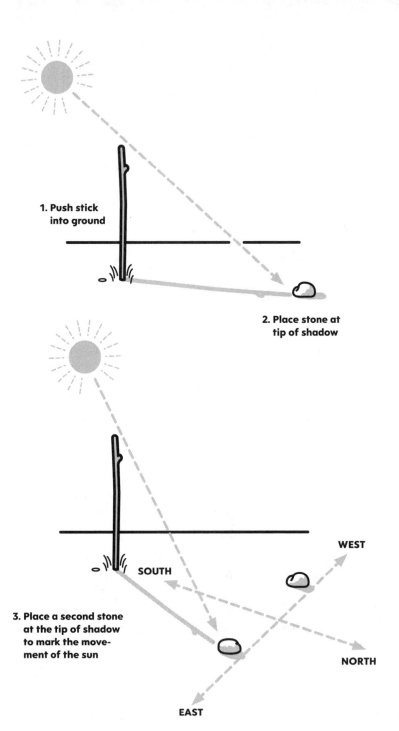

1. Push stick into ground

2. Place stone at tip of shadow

WEST

SOUTH

3. Place a second stone at the tip of shadow to mark the movement of the sun

NORTH

EAST

straight line bisecting the west-to-east line. (East is to the left of north.) Now you have a perfect "shadow" compass to point your way.

d. To remember the compass directions and how they move clockwise, think **N**aughty **E**mus **S**lurp **W**ine (i.e., North, East, South, West).

3. If you have an analog wristwatch with an hour and minute hand (in other words, not digital), use the watch to find your direction.

a. In the northern hemisphere, place the watch flat on the ground or in your hand and rotate the watch so the hour hand is pointing in the rough direction of the sun. (If it is Daylight Saving Time, point the previous hour at the sun. Example: If it is 3 p.m., point 2 in the direction of the sun.)

b. If it is morning (i.e., before noon), look at the hour hand and move clockwise to find the spot that is the halfway point between the hour hand and the number 12 on the watch. This halfway point is pointing south. (The opposite side of that imaginary line is pointing north.)

c. If it is after noon, look at the hour hand and move *counterclockwise* to find the spot that is the halfway point between the hour hand and the number 12 on the watch. The halfway point is pointing south. (The opposite side of that imaginary line is pointing north.)

d. In the southern hemisphere, point the 12 at the sun, not the hour hand. Follow the same instructions as above, only this time, the compass direction between the two hands will be north.

I like the wristwatch method because you can employ it on the go, without needing to stop moving and start planting sticks in the ground. However, the closer you are to the equator, or the higher the sun is in the sky, the more difficult it may be to pinpoint the sun's direction. In those cases, it is best to use the stick-and-shadow method.

NORTHERN HEMISPHERE

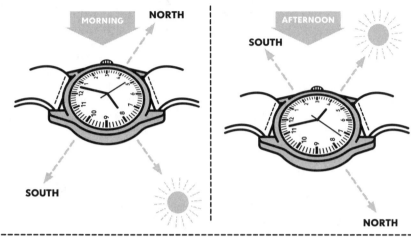

SOUTHERN HEMISPHERE

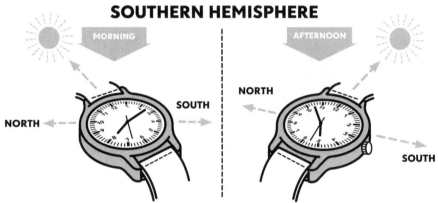

Choosing Your Survival Watch

The world of wristwatches feels quaint in an era when everyone can just peek at their mobile phone to tell the time. But smartphones and smartwatches both suffer from a fatal flaw: Their batteries need to be charged on a regular, often daily, basis. And they're infamously delicate devices that can't take a lot of wear and tear. That won't work if you're thrown into a survival situation.

Watches are classified by two main categories: quartz and mechanical.

A quartz watch is powered by piece of electrical circuitry run off a battery. (In some cases, the battery is solar-powered.) Quartz watch batteries are rated to last 2 or 3 years, often much, much longer. The "computer chips" that run these watches are so accurate that you can pretty much "set it and forget it." (A cheap $18 watch I discovered while writing this book loses only about 6 minutes in an entire year!) You can find watches that shrug off dampness after being dunked in 150 feet, 350 feet, or 650 feet (45, 107, or 199 meters) of water—and even deeper! Such watches may have tons of features that allow you to check the barometric pressure, someone's heart rate, sunrise and sunset times, etc. You probably don't want a lot of bells and whistles if you aren't committed to learning how to operate them all.

So: the ideal disaster watch . . .

1. **Is quartz**

2. **Is loaded with a fresh battery**

3. **Has a luminescent dial**

4. **Is built into a rugged case that can take a beating**

5. **Is water-resistant**

6. **Is attached to a comfortable band**

7. **Is easy to operate**

Believe me, many modern watches fit these criteria. (I list some brands in the resource section of this book on page 234.) You just need to get to a good watch shop or outdoor store and start asking questions. For most people, a solid "field watch" (for active outdoor use) or "dive watch" (worn by professional divers) is fine. Many of the disasters I discuss in this book will expose you to the elements and water, and these watches are designed to withstand these hazards.

You'll notice I didn't discuss mechanical watches. They keep time with gears and springs and a minuscule amount of oil to keep things ticking smoothly. Mechanicals are wound by hand, or wound automatically by the kinetic movement of your wrist as you wear them. Some of these watches are quite rugged and full-featured. But in an emergency situation, if you don't wind them or you take them off for a few days, the watch may wind down and the time may need to be reset. That's going to be hard to do if your mobile phone is damaged or its battery has run down, and you don't have a second accurate timekeeping device to refer to. That's the only reason I don't recommend mechanical watches for *most* people.

However, I will say that some military personnel prefer them. Why? Because, if they are ever in a situation where an "e-bomb" or radio signal has wiped out electromagnetic devices, mechanical watches will keep ticking. (In 2003, the U.S. Air Force detonated an experimental microwave device to take out Saddam Hussein's 24/7 TV news station. Or not. The tech is classified, and the Pentagon has never openly confirmed the existence of such a weapon.) In theory, such a weapon will kill quartz circuitry dead, not to mention a whole lot more.

HOW TO DRINK DIRTY WATER

You can live 3 days without water, but it's likely to be a miserable 3 days. Simply put, water is life. Your body is 75 percent water. When it's hot and you sweat, you lose some. When you use the bathroom, you lose some. You *even* lose some when the weather's cold. You lose a lot when you breathe. And you lose some when you're stressed. You must replace it in order to restore your body's ability to function. Typically, you'll know from your thirst level that you need to drink, but be prepared to hydrate even when you don't feel thirsty.

The ideal consumption is a gallon of water a day for each person. That's to drink, practice good hygiene, prep food, and wash dirty dishes. The bare minimum to sustain life is two quarts per person a day.

Some tips:

- Before an emergency strikes, be sure you have outfitted your home or bug-out bag with portable water filters, water purification pumps, iodine solution, or purification tablets. These are the classic tools of the trade.

- If you're stuck in a situation with limited water, start searching now to get more, so you can minimize your time without any. They have a saying in the military: "The best place to store water is your stomach." The point being: Never conserve your water to the point that you dehydrate.

- If you are stuck in your home, and you anticipate a water shortage or expect that the authorities will soon shut off the water supply, fill your bathtub immediately from the tap. That will last you quite a few days.

- In both urban and rural areas, if water is in short supply, government authorities and aid organizations will often truck in water. Ideally, authorities will distribute water in plastic bottles. But it's not a bad idea to take some empty containers when you go to distribution sites, just in case. You can always use the extra containers later to catch rainwater.

- In a natural setting, like forests or desert canyons, locate streams or rivers. That's the best because it's a) fresh water, and b) less contaminated because it's running. Resist the urge to scoop a cup up and guzzle it down. That's a great way to expose yourself to toxins, or microscopic parasites like *Giardia*.

- If you anticipate rain, take an immediate inventory of everything you have on your person, in your bug-out bag, or carrying in your party. Place everything you can use as a catchment device where it is out in the open, away from overhanging trees and other shelter. You can drink anything you catch in this manner as long as your catchment container is clean and the water fell directly from the sky, and never came in contact with other surfaces.

- For that matter, you can lick or sip rainwater off plant leaves after a good, hard rain. The catch: You must be absolutely certain that you have identified the plant as harmless. It would not be pleasant to lick poison ivy!

- After a heavy rain, you can dig a deep hole in mud and wait for it to refill from the ground-saturated water. Yes, it will be muddy, but you can collect it and use a water filter to clean it.

- If you are in a wintry environment, you can drink clean, fresh snow— but you must melt it first. Unmelted snow is far too cold for your body, and ingesting it may cause other health problems.

- Never drink saltwater or your own urine. Not even as a last resort. Besides being gross, their salt content is far too high for your system. You'll only want to drink more, and that's a path to diminishing returns. Don't go down it.

- Some commonly accepted ways to purify water found in the wild:

  A. Boil it for 1 full minute, adding 1 minute for each 1,000 feet (305 meters) you are above sea level. Don't exceed 10 minutes or you'll be evaporating too much. Let it cool, pour it into your water containers, or use the hot water for soup, tea, or to cook dehydrated meals.

B. You can drip 2.5 percent iodine solution with an eyedropper into your container of water. Eight drops per quart is sufficient. Swish or shake your container. Wait 15 minutes before consuming.

C. You can also use water purification tablets. Typically you need to strain or filter out the obvious, visible gunk in the water, then drop a purification tablet into the container, mix for about 10 minutes, and wait about 30 minutes before you drink the water. The chemical in these tablets is *usually* iodine or some form of chlorine, but that varies greatly by manufacturer. Follow the instructions given on the package of the product you're using. Tip: Choose tablets with the active ingredient *chlorine dioxide*. They neutralize the largest number of contaminants.

D. If you have a water purification pump, you can literally pump water straight from a pond, and pass it through the filter and into your water bottle. If you don't have such a pump, you can use a "straw" device that works the same way. In a notebook, keep track of how much water you have filtered cumulatively with each device. Most manufacturers tell you how much water you can filter before the device or filter needs to be replaced. You can certainly keep a few extras on hand.

KNOW YOUR TOXIC PLANTS

"Leaves of three, leave them be," goes the old saying, and it's correct—sort of. But that maxim really only applies to poison ivy (*Toxicodendron radicans*), which *always* grows in a very distinctive three-leaf configuration. Poison ivy's relations—poison sumac (*Toxicodendron vernix*) and poison oak (*Toxicodendron diversilobum*)—sometimes grow in leaves of three, but often don't. We say these plants are "poisonous" because their leaves contain urushiol, an oily sap that binds to human skin and causes irritating blisters, rashes, and sores that will drive you crazy with itching for weeks.

But there are plants that make poison ivy look like a quaint trip to the arboretum. I'm talking about plants whose toxins will destroy your heart, wreak havoc with your nervous system, or shut down your liver permanently. You could spend your life studying killer plants, and people certainly have. My chart outlines just a few of the dangerous plants you'll find in North America. If you're going to spend a lot of time in the outdoors, you owe it to yourself to get a good plant book and bone up on what you might encounter. As a general rule, you should never touch, eat, or burn a plant that you can't identify on sight. (Dead, dry wood is usually fine to burn.)

I wish I could say that dangerous plants lurk only in the wild. But they're as close as your backyard. Some of the loveliest flowers grown for ornamentation will sicken, harm, or even kill pets, children, and adults—if strange, invasive weeds on the edges of your property don't kill you first.

Just to keep things unpredictable, nature is forever changing up its toxic brew. In semi-arid Colorado, where I live, rainfall impacts the toxicity of certain plants. Jimson weed (*Datura stramonium*) grows like . . . well, a weed on our highways. Kids who've heard that the plant has hallucinogenic properties will often collect the seeds from its dried pods, boil 'em, and swallow 'em. In years of drought, one seed might carry the payload of three seeds in a rainy season! Kids expecting a high often ended up in my ER instead, treated for severe damage to their nervous system.

One plant on my list—water hemlock (genus *Cicuta*)—is probably the deadliest in North America. If you saw it growing out of a riverbank, it would remind you of baby's breath in a flower arrangement or Queen Anne's lace. It would dazzle you with its lacy "umbrella" of tiny white flowers, humming and buzzing with pollinators. But it'll kill you dead—violently, with terrifying convulsions—in under an hour if you ever ingested a bit of its toxin.

Serious gardeners know to wear garden gloves when handling most plants in their gorgeous beds. They minimize how much they handle a particular plant. And they're careful to wash up afterward.

But in the wild, I suggest a different approach: Look, don't touch.

Especially mushrooms. Even experienced mushroom collectors make deadly mistakes sometimes. There's an old saying: "There are old mushroom collectors. And there are bold mushroom collectors. But there are no old, bold mushroom collectors."

Be careful out there.

The Urushiol Trio

These are the classic three known for causing skin rashes. If exposed, wash quickly and be careful handling clothes, shoes, or even pets that may have been exposed.

Poison ivy (*Toxicodendron radicans*)

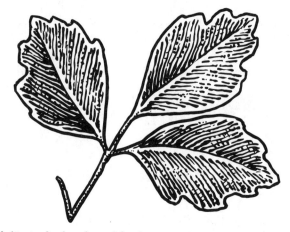

Poison oak (*Toxicodendron diversilobum*)

Poison sumac (*Toxicodendron vernix*)

Pretty (deadly) garden flowers:

You'll find these beauties in almost every serious garden. Castor bean plant is famous for the "spy" poison made from its bean. Oleander, for example, is often used as a hedge in public places such as schoolyards. All of these plants contain toxins.

Castor bean plant (*Ricinus communis*)

Daffodil (genus *Narcissus*)

Larkspur (genus *Delphinium*)

Foxglove (genus *Digitalis*)

DR. DISASTER'S GUIDE TO SURVIVING EVERYTHING

Monkshood (genus *Aconitum*)

Oleander (*Nerium oleander*)

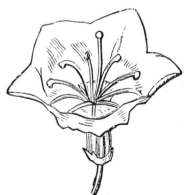

Mountain laurel (*Kalmia latifolia*)

The Deadly Weed Growing in Your Backyard

Tall, fleshy plant with green leaves, freakishly pink clusters studded with black berries that can attract children. Remove 'em when you see 'em.

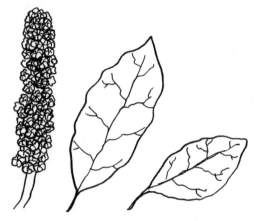

Pokeweed (*Phytolacca americana*)

Ouch, It Pinches!

Each leaf is studded with tiny "needles" that break off when they come in contact with skin, and inject histamine and other toxins, causing a stinging sensation.

Stinging nettle (*Urtica dioica*)

Deadliest Plant in North America

A member of the family that includes carrot, parsley, and celery—only, this guy makes you sick in 6 minutes, dead in 60. In survival training, soldiers are trained to never eat any plants that have lacy flowers. This plant is why.

Water hemlock (genus *Cicuta*)

YOUR FAMILY'S RENDEZVOUS PLAN

It would be nice if disasters happened when we and our family members were all together in one place. That way, we could have a quick conversation with everyone about the best course of action. But the world doesn't work that way.

The events of 9/11 happened early on a Tuesday, at a time in the day when most New Yorkers were at work. In the ensuing chaos that enveloped the Tri-State area (New York/New Jersey/Connecticut), landline phone service and cellular service were saturated, and many people were unable to communicate with their friends, family, and loved ones.

Such a situation is a recipe for confusion and anxiety, to say the least. These days, we are accustomed to being able to pull a phone out of our pockets, and calling or texting the people closest to us.

But what if you can't? What if you're separated by a disaster, and you have no way to connect with family? How do you find them? How do you reconnect?

Well, believe it or not, the military grapples with this logistical quandary all the time. We teach soldiers techniques that help them problem-solve their way out of almost anything, including being separated from their unit.

As with so many things in the military, this approach is best understood by an acronym: PACE. This stands for Primary, Alternate, Contingency, Emergency.

Say your **Primary** plan for reconnecting with your squadron is using a company Jeep to get to a previously determined rendezvous point.

What if the Jeep breaks down? No problem. The **Alternate** plan is to borrow another vehicle.

No vehicle in sight to borrow? The **Contingency** plan is to steal a civilian vehicle. (Notice: The situation has escalated, and we're now forced to take a more drastic measure.)

What if the civilian vehicle breaks down en route to the destination point? No problem. The **Emergency** plan is walk.

To boil down the message of PACE, it's not enough to have a Plan B. You need a Plan C and D as well.

> ## IT'S NOT ENOUGH TO HAVE A PLAN B. YOU NEED A PLAN C AND D AS WELL.

How do you decide your rendezvous point? Choose locations that everyone in your family knows. For years when our kids were young, I would tell them that in case of an emergency, they were to figure out a way come home to our house, even if they had to walk. If I or their mother could not be reached, they were to go to their aunt's house in another town. That was as extensive as I could make our family plan, considering the age of my kids at the time.

The best rendezvous sites are places that can be accessed at all hours, places that are easy to locate, places that have some significance for your family. Telling the family to meet you at your place of work is not a good option, because that location may be only accessible during office hours.

If I were traveling with my family to a city like New York for a few days, I'd remind everyone that if landlines are down and they can't get a signal on their mobile phones, they should always try to send a text. Very often texts will send even if a call cannot go through. Failing that, I might outline the following plan:

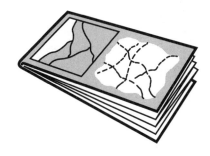

"If we get separated and can't communicate, meet back at the hotel. If the hotel is not an option, I'll be in front of the Empire State Building every day from 3 p.m. to 6 p.m. If that's not possible, we'll meet at Uncle Will's house in New Jersey, just over the George Washington Bridge. If that's not possible, we'll all meet at the family cabin at the lake in upstate New York."

It is unpleasant to contemplate a scenario in which your family would be forced to move to Plan D. But many disasters—massive floods, blizzards, tornadoes, man-made attacks on cities—do indeed impact vast geographical regions.

You need to strike a balance between hammering out a plan that is realistic and achievable for your family. If you have children, you will need to factor in their ages, temperaments, and maturity. You don't want to frighten them unnecessarily, but you want them to have a solid idea where to find you if they cannot easily pick up a phone or text, and if your home or neighborhood is somehow not an option.

ZOMBIES

I venture to say that most Americans have never taken the time to peruse the blog at the website of the U.S. Centers for Disease Control and Prevention in Atlanta. I can't say that I blame them. The agency's releases are critically important, but typically very dry material. But in spring 2011, the CDC used that blog to make what is arguably the most out-there post in the history of the federal government.

Writing on behalf of the agency he ran, then Assistant Surgeon General—Rear Admiral Ali S. Khan—wrote these words: "There are all kinds of emergencies out there that we can prepare for. Take a zombie apocalypse, for example. That's right, I said z-o-m-b-i-e a-p-o-c-a-l-y-p-s-e. You may laugh now, but when it happens, you'll be happy you read this."

Khan then went on to list a slew of action steps wary Americans could take to protect themselves and their loved ones from the shambling undead. Stuff like keep a gallon of drinking water per person handy, along with a bottle of bleach in case you need to purify water or sanitize objects that have come in contact with the putrefied flesh of the not-quite-dead. It was downright hilarious, money-in-the-bank comedy, when you consider the source was a federal agency.

> Plan your evacuation route. When zombies are hungry, they won't stop until they get food (i.e., brains), which means you need to get out of town fast! Plan where you would go and multiple routes you would take ahead of time so that the flesh eaters don't have a chance! This is also helpful when natural disasters strike, and you have to take shelter fast.

It was a stroke of genius, really. The post came a month before hurricane season officially starts in the United States, and the agency was looking for a way to galvanize Americans into taking disaster preparedness seriously. As the spokesperson who dreamed up the blog post later told reporters, "I don't have to tell you that preparedness and public health are not the sexiest topics."

How well did it work? Normally that website gets between 1,000 and 3,000 hits a day. The post went up on a Monday. By that night, 30,000 visitors had tuned in to check out the post.

By Wednesday, the server crashed.

The CDC followed up the post by hosting a video contest to create a Zombie Apocalypse public service announcement ad. And today, if you go to their website, you'll find an entire package of zombie material in English and Spanish, including curriculum plans, a graphic novel, and spooky posters. It's incredibly detailed stuff, like asking children to plan escape routes, prearranged meeting points, first-aid supplies, and so on.

Why am telling you this? Is it because I think we really are in imminent danger of an attack from the shuffling undead?

Nope. I'm telling you about it because this is a perfect way to get your kids interested in disaster preparedness in a way that is not alarming or frightening.

Well, okay, I suppose zombies are pretty scary. But not in the same way as some of the real-life disasters in this book are. Kids know zombies are scary, but they also know that zombies safely dwell in the world of make-believe.

Everything contained in the teaching materials are things kids need to know in case of any emergency, not just a zombie attack.

Look for this free material at the agency's website: https://www.cdc.gov /cpr/zombie/index.htm

And just one word of advice. Don't read it after dark.

PART II

NATURAL PERILS

ANIMAL ATTACKS

When I was still in medical school, the state of New Mexico experienced an outbreak of hantavirus. That's a terrible disease that humans catch from inhaling dried, aerosolized urine and droppings of rodents. I know. It sounds like the grossest, most outrageous thing in the world. Who's going around inhaling dried mouse urine? But that's exactly the kind of disease that can spring up out of nowhere in a city located in the middle of the desert.

Next thing you know, the local news guys were looking for doctors to comment on the outbreak and came to the ER on my shift. An emergency room doctor has to be quick, steady, and fast on his or her feet in order to break difficult news to patients *fast,* and give them exactly the information they need to make a critical decision about what care they will consent to. On camera, I imagined that I was speaking to one of my patients. What did he or she need to know? If your mother, father, sister, spouse, or child were stricken with something, what would *you* need to know *right now* to be able to help *me* do my job? Whatever I said on camera that day must have worked because soon after, the crew came back again and again.

About the umpteenth time a reporter doing standup stuck a mic in my face, I said, "Look, I have work to do. Don't you guys have a doctor who can help you with this kind of stuff?"

They cut the camera immediately. One of the producers came over and confided, "Well, as a matter of fact, no, we don't. You wanna do it?"

I guess you could say I owe my news career to mouse pee.

But seriously, the number of animal species that share our world is seemingly endless. It pays to know how you will react if you or a loved one have a disastrous encounter with a domestic or wild animal.

Wild animals are truly amazing, but should always be viewed from a distance. Ninety percent of the time wild animals will act predictably because, like us, their flight-or-fight behavior is hardwired. At the first sight of a chattering, bipedal creature wearing a backpack, they'll flee. The trouble starts when they feel cornered, surprised, or have young to protect.

At the other end of the spectrum are domesticated animals such as pets and livestock. Dogs and cats share our homes, snuggle with us, and may even feel protective of us. Livestock, at the very least, are accustomed to human presence and interaction. They know that our approach usually means food is coming, so they tolerate our strange behavior. But you should never take any of these generalizations for granted. A dog or cat you don't know personally or very well can act unpredictably and cause you harm.

What to watch out for...

Rabid Animals

Rabies is a horrific virus (*Rabies lyssavirus*) that passes to humans from the saliva of animals. It devastates nerve cells, causing fever, headaches, delirium, and ultimately death. The scary thing about rabies? Once symptoms start, they're *100 percent* fatal. Only two people in the history of the United States have survived rabies, and the list of survivors worldwide remains under twenty. About 59,000 people a year die from rabies worldwide, which is appallingly tragic.

Unlike pets, humans don't get vaccinated for rabies. A so-called pre-exposure vaccine is available, but it's typically reserved for lab workers who study rabies itself, or people who work regularly with wildlife.

Once bitten, you only have 7 days before you become symptomatic. Ideally, you must reach a doctor or medical facility no later than 3 days to receive treatment and have any chance of survival. For this reason alone, I would never embark on a wilderness vacation unless I was certain of reaching civilization within this window of time.

You may have heard stories about rabies patients receiving painful stomach injections. That doesn't happen any more. (In fact, it never really did. Back in the day, patients received those shots subcutaneously, in whatever fat they carried in their midsections.) Today patients receive four separate injections—on the first day they see the doctor, and then on the third, seventh, and fourteenth day after that. Adults get the shot in their upper arm. Kids get it in their thigh muscles. When you were a kid, you were probably told to beware of rabies following a bite from a dog or a rat. But the biggest rabies carriers in the United States are bats, skunks, raccoons, and foxes. There's never been a documented case of rodent-to-human rabies transmission in the United States.

Doctors like me have such faith in this statistic that if a patient arrives in an emergency room complaining of a rat bite, we immediately rule out the chance of rabies. We make the same determination in the case of a rabbit bite, because for some reason rabbits, hares, and other lagomorphs don't appear to transmit rabies. Equally surprising, some of the most unlikely animals, such as cows, can transmit rabies.

But you're probably not going to be bitten by Bessie. The Centers for Disease Control and Prevention logs about 5,000 cases a year of animals identified as carrying rabies. Ninety percent of them are *wild* animals, not pets or domesticated animals.

How do people get bitten? They startle bats or raccoons in abandoned barns. They try to shoo a frightened bat out of their tent when camping. They're bitten by feral dogs or cats they're trying to comfort or feed. They try to rescue a forlorn creature that looks sick or hurt.

Here's the good news: In the United States, most dogs and cats are vaccinated for rabies shortly after birth or adoption. This completely eliminates their chances of carrying the virus. But this is also why stray dogs and cats should never be approached by anyone other than an animal control officer. You simply do not know if those strays are carriers. The old assumption that a rabid animal is foaming at the mouth is unreliable. If a supposed domestic animal is behaving strangely, avoid them. If you see a nocturnal animal like a raccoon in the daytime, avoid it. You don't ever want to get close enough to an animal to see if it's foaming at the mouth.

Here's the catch: When traveling outside the United States, you must expect that people are less likely to vaccinate dogs and cats in, say, Mogadishu than in Milwaukee. If you are bitten by a domesticated animal in a foreign country, you cannot easily rule out rabies. Wash the wound with soap and water, cover it, and get to a doctor. Be prepared to share all pertinent details of the incident with your medical provider.

Cats and Dogs

In general, it's a wise policy to approach any pet you don't know with caution. Many people have had the experience of nicely asking a dog owner if their pooch is friendly, getting the thumbs-up from the doting human, and recoiling in horror when Bouncer proceeded to take a chunk out of their forearm. You can't always trust the human who spends their every waking moment with a particular animal. They may be oblivious to the truth, or singled out by the animal for loving treatment.

If you're bitten by a dog and can safely rule out rabies, treat it like any other cut. Wash it immediately, pat it dry, cover the wound with a bandage, and monitor it for infection. Cats, and sometimes dogs, can carry *Pasteurella*, a bacteria that can cause swelling, infections, or more.

Cats are more problematic. Their fangs and claws are sharper, more needle-like. They penetrate deeply into skin, and can damage a tendon or blood vessel. I've seen cases where a piece of a cat's fang broke off, lodged in a person's hand, and had to be surgically removed.

In my ER days, I also became accustomed to meeting sweet elderly women who arrived at the hospital with very similar ailments. Invariably, they lived alone. One or both of their hands were red, painfully immobile, and big as a balloon. The patients sometimes suffered from swollen lymph nodes, aches and pains all over their bodies, fever, and abdominal pains.

The diagnosis: cat-scratch fever. The microscopic culprit was *Bartonella henselae*, a bacterium carried by pet cats. It was always the same story: The sweet little kitty had crawled under the bed, and when the owner reached under to coax Felix out, the cat not-so-playfully mauled their human's hand. The owner never thought much of it at the time. After all, it was just a scratch! But days later they were downing a course of antibiotics to get back to normal.

It's a warning for all of us never to shrug off an encounter with a dog or cat. A feral animal is always a danger, but even a pampered house pet can pose threats.

Treat the scratch or bite—but watch it!

Rats

Did you think you were going to get off easy on rats? No such luck. These large rodents carry a number of pathogens, but probably the most common concern is rat-bite fever (RBF), chiefly transmitted to humans from the animal's saliva, but you can also get it from food and water contaminated by their urine and droppings. It's caused by two different bacteria: *Spirillum minus* in Asia, *Streptobacillus moniliformis* in North America. Not every bite will get you sick, but if the rat is a carrier, infected people will start showing symptoms in 3 to 10 days. Fever. Chills. Joint Pain. Nausea and vomiting. All the usual gunky stuff, plus a rash at the bite site that turns into a throbbing, ulcerating wound. It's highly treatable with antibiotics, however, as long as you get to a doctor. Unlike mice, rats are bolder about defending their turf, and can be aggressive, not easily shooing when a human draws near. The best defense is staying on your guard and avoiding them like the plague.

Speaking of which, the nastiest, most medieval of diseases—the plague (caused by the bacterium *Yersinia pestis*)—still exists in the world. In the United States, it survives predominantly in the Four Corners region of the Southwest. So if you're traveling anywhere in Utah, Colorado, New Mexico, or Arizona, exercise caution. The disease is spread by fleas that feed on the blood of rodents and squirrels, and then hitchhike on the bodies of other animals, including pets. If you're hiking in a canyon, for example, and spot a brilliantly colored feather lying on the ground, skip the impulse to collect this gorgeous natural artifact. Same goes for any dead birds or animals, no matter how interesting its pelt, feathers, or skull.

Bears

One of the most famous entries in the logbooks of Lewis and Clark focuses on their encounter with a grizzly bear. Some members of the explorers' entourage stepped off a raft while floating down a river and attempted to pick off the bear, thinking this was an easy kill. But the first few shots only enraged the bear, and the fleeing men needed much more firepower to dispatch their attacker. After that, Lewis and Clark laid down a new rule: No one gets off the raft to shoot at bears, ever.

Bears are found on four continents in the world, but the two you are most likely to encounter in North America are the black bear (*Ursus americanus*) and the brown, or grizzly, bear (*Ursus arctos*). The animal's color is not always accurate. A black bear might be brown in appearance, and vice versa. Whatever the species or subspecies, they're magnificent creatures but their survival is increasingly under pressure by the loss of habitat and their contact with humans. Hikers traveling on foot in bear territory sometimes wear bells to announce their presence, but the research on their utility is unclear. You probably ought to be making a lot more noise as you hike to ensure that you don't surprise bears and other animals, but this understandably demolishes the wilderness experience most people are seeking.

- Never feed bears. Employ smart food and trash practices when in bear country. Human–bear contract is usually initiated when the bear wanders into a human neighborhood or campsite in search of food. Campsites in bear territory typically broadcast bear safety tips and will be equipped with bear-resistant trash cans, which you should utilize to dispose of your trash. Campsites will also indicate exactly how they prefer you store your food while in the area. Sometimes they insist on bear-resistant canisters, which suppress food odors and are hard for bears to open. Other times, you'll be instructed to suspend your food in a special bag high off the ground from a tree limb, or in your locked vehicle and out of sight. Yes, *locked*. Bears have been known to open unlocked car doors.

- Observe bears from a safe distance. Sighting a bear does not always mean trouble. In most cases, if the animal does not know you're there, you can enjoy the experience. Snap photos if you wish, but don't go any closer and don't reposition yourself to get a better shot. Be on your guard and be prepared to leave the area.

- If bears hear or smell humans, they will most likely leave the area to avoid an encounter. Occasionally, your sudden appearance will surprise a bear. If that happens, don't run away. That's the one thing you absolutely never want to do. That will only stimulate their instinct to chase. Stand your ground. Clap your hands. Wave your arms in the air. If you have a windbreaker or jacket, open it to appear larger. (Moving to a higher position can have the same effect, but don't box yourself in by, say, climbing a tall rock you cannot easily escape.) Avoid climbing trees for the same reason. Speak to the bear in calming tones. Do not scream. Do not allow small children to shriek. Leave the area as soon as you can. Don't turn your back on the bear. You want to keep your eyes on them at all times. Moving sideways is often a better bet because you'll leave their sight quicker. But this is not always possible in wooded environments. You don't want to leave the trail and create a worse problem by losing your bearings and getting lost. The next-best option is to walk carefully backwards in the direction that you came, until you are a safe distance away.

- Most bears simply want to check you out, maybe huff and puff, then skedaddle. If the bear stands, they're just trying to get a better look at you. It does not mean an attack is imminent. Same goes if they make woofing or chomping noises. They're trying to get you to leave them alone.

- Never get between a mother bear and her cubs. Mother bears are under enormous stress, because their daily routine not only requires them to find enough calories for themselves to consume, but enough for each one of their young. Without thinking, they will put their own lives in danger to protect one of their cubs. If you see a large bear, never rule out the possibility that this individual is a female traveling with cubs.

- If you are walking backwards and the bear follows you, stop. Stand your ground. Continue your waving, clapping, and calm speech. When the bear stops moving, move away. Now is a good time to think about what you might use to fend off an attack if one occurs. Keep your backpack on; you might need it to protect your back.

- If a *black bear* (black fur, as much as 500 pounds or 227 kilograms) attacks, your best option is to fight it off using whatever tool or object you have handy. Focus your blows on the animal's face. Blows directed at their enormous bodies will not have much effect. Do not play dead. You have a better chance of chasing this species off if you fight.

- If a *grizzly* (brown fur, hump on back, up to 1,000 pounds or 454 kilograms) attacks, drop facedown, backpack up, and play dead. Spreading your legs will make it harder for the bear to flip you over. Wait out the attack. Terrible as it sounds, I've heard some experts say you are better off letting these behemoths chew your arm or thigh muscle until they get bored than trying to fight them off. Other guides say it's better to fight back, focusing blows on the animal's face, if the encounter turns violent. Sad to say, there are no guarantees.

- Bear spray is legal in all fifty U.S. states. This is a non-lethal, capsicum-based (i.e., hot pepper) deterrent that you can use to discharge a very large, diffuse cloud between yourself and a bear. The chemical compound is not nearly as strong as the pepper spray you would use to ward off an attack by a human. Don't confuse the two, and never use human pepper spray on a bear. The stream of the human product will be too narrowly focused to do the job. (Some cans of human pepper spray are designed to use in really tight quarters and to hit the attacker directly in the eyes.) You don't have to be an expert shot with bear spray. The whole point is to put a giant, irritating cloud between you and the bear. The bear barrels into the cloud, detects the offensive odor, stops its advance, and leaves. Bear spray should be discharged when the bear is still about 30 feet (9 meters) away, so you still have a sizable distance between you.

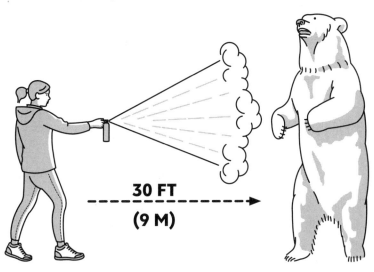

**30 FT
(9 M)**

This is all nice in theory, but I should mention that not all experts agree on the efficacy of bear spray. For that reason, buy only the best product possible. Carry it on your hip in a holster, not in your backpack or luggage. Practice with an empty can so you can get the hang of removing the spray from your holster, removing the safety clip, getting a feel for the correct distance, and shooting. (The best manufacturers sell practice cans.) Never apply bear spray to your body, your tent, your backpack, or your camping equipment; it doesn't work that way. In some large campgrounds, you can rent bear spray canisters and return them when you get back from what we all hope is a perfectly uneventful excursion.

If you are traveling to a foreign country, check local laws to see if bear spray is lawful. Bear spray is not permitted on U.S. flights in any form of luggage. All this aside, you should never allow having the spray lull you into a false sense of security. Animals as a rule are completely unpredictable. Shockingly, some bears have not read the fine print on the can of bear spray! Your best bet is to avoid bear encounters altogether, and to follow all the steps I've outlined to put distance between yourself and a bear before it becomes necessary to use spray. You never want to find out if bear spray works.

Snakes

One of the most outrageous stories of 2020 was the tale of a Queensland, Australia, man who was stopped by cops after driving recklessly down the highway. Turns out, he was using a knife and his seat belt to fight off a snake that had somehow managed to crawl into the cab of his pickup truck. This was no ordinary snake. It was an eastern brown—the second-most-venomous snake on the planet and the species responsible for the largest number of human fatalities Down Under. Luckily, the man was unbitten, and the snake—discovered in the back seat of the truck—was dead.

Luckily, this story is an aberration, and unlikely to happen to most people. Snakes are deaf and don't see very well. They rely on vibrations transmitted through their bones and skull for "hearing," and they sense prey with a high ability to detect infrared radiation, i.e., body heat. In most cases, the simple act of walking or hiking in their territory will send them scattering, but you really must be attentive when in a wilderness area. A snake will feel you coming if it's on the ground, but might very well miss you if lounging in or dangling from a tree.

Ten years ago, doctors and animal experts advised people to suck out the venom if they were bitten by a snake. Nowadays we don't. First, people make the wound worse by cutting too deep, and hitting nerves, tendons, or important vessels. Second, it's unlikely that you'll get much of the venom out; it will have already begun flowing through your system. Third, sucking it by mouth risks further self-infection, or infection of your rescuer, since we all have abrasions in our mouths. You'll find snakebite kits on the market offering a knife and a suction-cup device to use—but experts don't recommend using them.

I know it's hard to tell someone this, but if you are bitten, you must really try to keep calm. A faster heart rate circulates the poison faster.

Have the person who has been bitten remove all rings, watches, or bracelets—i.e., things that might constrict blood flow if a hand or wrist becomes swollen. Keep the wound as low as possible. If ice is available, ice the wound. Move carefully and deliberately toward your transportation.

Victims often have more time than they think to make it to a hospital. Adult snakes are capable of "dosing" their venom; it's possible you may not have gotten too large an infusion. Baby snakes can't control their venom and are thus more dangerous.

As with bears, the best protection is avoiding an encounter in the first place. In an ideal world, you'll be able to tell your ER docs the species of the snake that bit you so they can treat you with the correct antivenin. There are two types of venom: ones that attack your nerves, others that attack your blood. Doctors tend to know which species are in their region, and if you cannot ID the species, they will make their treatment decisions based on your symptoms.

WILDFIRE

Waldo Canyon was once the jewel of our local Colorado wilderness. A hiker's dream. Pristine views that allowed you to see Pike's Peak, the 14,115-foot (4,302-meter) neighboring mountain. When the wildfires started in June 2012, all I kept hearing on the news was that the flames would *never* reach Colorado Springs. Yes, the fires were only 3 miles (4.8 kilometers) away as the crow flies, but our city was protected by the vast expanse of nearby Queens Canyon.

A friend of mine who is an avid mountain biker told me with absolute conviction in his voice, "John, I know every inch of Queens Canyon, and there's no way fire is going to cross *that*."

My buddy's assessment aligned with that of the experts interviewed by the TV station I was working with in those days. There was just no way, the experts said, that the fire could jump the rocky canyon and reach our homes. The fire would burn itself out before it reached downtown.

Shows you how much experts know.

In a freakish turn of events, a thunderstorm whipped through as the fire raged out of control. Murphy's Law went into effect. The rain was not enough to put out the fires. But the storm's resulting 50- to 60-mile-per-hour (80- to 96-kilometer-per-hour) winds caused the fire to do the "impossible"—jump the canyon and blow across the tops of the canyon's trees.

I was at the TV station that afternoon, watching the news reports come in. I grabbed the phone and called my mountain biker friend. "You're not gonna believe this," I said. "The fires crossed the canyon. They're evacuating the city!"

Utter silence on the other end of the line. He was flabbergasted.

If we had stayed in town, we'd have been toast. It took firefighters 18 days to control the blaze. By then, the fires had destroyed nearly 350 homes and caused $450 million in damage. It became the most destructive fire in Colorado history to that date.

Wildfires are normal in nature. Lightning hits a tree, a spark ignites dead vegetation, and the resulting fire consumes what it can in the area until it runs out of fuel. But then humans enter the picture. We build homes close to forested areas. We build fires to cook our food while camping. Or we suppress natural fires when they occur, so the dead vegetation builds up in a region, slowly becoming a tinderbox that is ready to explode. Suddenly, the natural wildfire is made many times worse.

The big shocker? People don't realize that in a wildfire situation, the authorities will literally triage an entire region. By this I mean that experts will look at an entire neighborhood and say, "Okay, all these houses have wooden shingles on them and they're surrounded by juniper trees, which is a highly flammable species. There's nothing we can do to save this neighborhood. It would be a waste of resources. Let's move on."

You read that right: Those charged with containing the blaze might just sacrifice an entire area for the greater good of managing a whole region. This means that you cannot bank on first responders to save your home—*or your family*—if you do not evacuate. They cannot waste that precious time and effort. They are too busy focusing on the neighborhoods that they *can* save.

So: If you evacuate in a wildfire situation, you must be prepared for the possibility of losing everything to the disaster. That's why having a bug-out bag, checklist, and family travel plan ready is so crucial (see pages 51–60).

When planning a hike or camping trip to a wilderness area, you must always check the fire danger level. This is an assessment of how easily a fire can start, given current weather conditions, recent moisture levels, and local climate conditions. You will generally see signs posted near the entrances to state and national parks stating the current daily conditions.

In the United States, the National Fire Danger Rating System (NFDRS) has five levels, from LOW ("fuels do not ignite easily from small embers...") to EXTREME ("fires of all types start quickly and burn intensely..."). The signage announcing these levels are typically marked by everyone's favorite ursine ranger, Smokey Bear.

In private campsites, whoever is renting you the space will instruct you on how they recommend you start, control, and extinguish your campfire or grill. If you're embarking into a true wilderness area, it's your responsibility to educate yourself about the current threat level and stay on top of the news via your mobile device or crank radio. This might feel like an intrusion upon your lovely wilderness experience, but you don't want to wake up one morning to find that you are surrounded by flames. Your first-aid kit on these adventures should be outfitted with N95 or P100 respirator masks, goggles, gauze, tape, and burn creams and gels. And as you make your way into the wilderness, give an occasional thought to how you will depart the area if you have to.

Many times a wildfire is raging many miles from your home, authorities have not evacuated the region, but you are still impacted because the air quality has degraded. Everyone—adults, children, seniors, and pets—may have trouble breathing. Stay indoors, keep windows fastened, and use air conditioning in summer. A single-room air cleaner with a HEPA filter can help remove some of the particles from the air. Refrain from exercise during this period. You should only use a respirator if you are running errands outdoors, but some people with existing respiratory issues might need to wear them indoors.

If the fire becomes worse, the only smart choice for survival is evacuation. Yes, there are some basic tips for staying alive if you're trapped in a wildfire. We've all heard amazing stories about people who survived fires by jumping into their pool. But you don't want to be caught in that situation. Why? The biggest killer is smoke inhalation. Wildfires heat the air to temperatures that become unbreathable. The oxygen is polluted by smoke, ash, embers, and toxic gases. Heat in your airways damages sensitive tissues, causing your throat to swell shut. Humans cannot survive long in such an environment. Making matters worse, fires are rapacious in their hunger to feed themselves. They move a heck of a lot faster than humans can run, and the front can easily blow over to where you've taken shelter.

Some scenarios:

- If you're trying to evacuate in your vehicle, shut off your fan/AC, or recirculate the air already the car. Keep your windows shut. Automobile tires are known to deflate if the surrounding air becomes too hot. If you're trapped, try to move the vehicle to a location that is free from as much vegetation as possible. You do not want to be close to any combustible material, including any wood-frame structures. Shut off the motor. Call 911 and give them your location. Slide off the car seats and get as low as possible. If you have water, wet a jacket, shirt, or blanket, and cover yourself with it. Use a respirator mask if you have one. Wait for help, or for the fire to pass you by.

- If you're in your home or a place of work, assess the structure. If it's covered in wood, consider moving to another location in the neighborhood. If that's not possible, consider using some of your precious time spraying water over the sides and roof of the structure. Fill sinks or tubs with water. Wet washcloths, towels, and other fabrics for everyone in your party. If you have blankets, jackets, or any other large fabric items, wet them as well. Notify authorities of your location. Keep your windows shut. Unlock your doors so rescuers don't have to waste time breaking down the door when they come. Move to a room that has the least amount of combustible material. (Strip curtains, fabrics, and rugs away so the fire does not have a way to encroach upon your location.) Cover yourself with the larger items like blankets and jackets. If the smoke gets bad, use the smaller wet items to cover your face and shield your lungs from inhaling smoke. Hunker down, and wait it out.

- If you're stuck in a forested area, you need to draw from both of the previous scenarios. Move to an area of sparse vegetation. Look for a ditch or lower elevation where you can hunker down and take shelter. If you have time, you can dig a trench. Try to put a body of water between you and the oncoming fire. Wet your clothing and other fabrics you have. Lie facedown and cover yourself with your wet garments or blankets and wait for the fire to blow over, or pass you by. If fabrics are not an option, you can immerse yourself in the water or mud.

EARTHQUAKES

Many of the disasters in this book can be predicted ahead of time. We always know when a hurricane is coming, and when floods are likely. Wildfires might spring up in a flash, but we usually know where they are and how fast they're moving. Even tornadoes can be predicted based on current weather conditions.

But there's no meteorologist or seismologist alive that can tell you exactly when an earthquake will occur. The best guidance you're ever going to have is knowing if you are living, working, or traveling in an earthquake zone. In many countries, your chances of survival in an urban setting really depend on how well the structures are built. If you're traveling in an earthquake-prone region or nation, it's smart to think about the sturdiness of every structure you enter, and plan accordingly.

In 1997, the two earthquakes that struck near Assisi, Italy—the city famous as the hometown of St. Francis—were "only" 5.5 and 5.7 magnitude on the Richter scale. Some major metropolitan cities in the world would have been able to withstand such a blow with little damage, because they're filled with modern structures that are designed to be as earthquake-proof as possible. But Assisi, a city of 5,000 that dates back thousands of years, was filled with beautiful old buildings constructed the old-fashioned way, as they have been for centuries, from unreinforced rock and mortar. It's no wonder the roof of the St. Francis Basilica collapsed.

In short, you really don't know what's in store for you in an earthquake until it happens. Aside from earthquake-proofing your home or office, and arranging a rendezvous plan with your family, there is little you can do to prepare. You could, for example, pack a small first-aid kit or go-bag to keep with you in the car, office, or home, but there's no guarantee you will have access to it when the temblor hits. Rather than focus on having the right gear, focus on understanding the types of shelter that are most likely to keep you safe.

Much of what we've been told about surviving earthquakes is antiquated. For example, standing in a doorway to ride out an earthquake was great advice when thresholds were the sturdiest part of a home. These days, the construction in most houses is of equal stability throughout the structure. A threshold is no better or worse than the rest of the structure. Outside of new buildings in known earthquake zones, many structures are not earthquake-proof or even earthquake-resistant. If you rely on the old doorway advice, you risk death.

Emergency rooms see three waves of victims after quakes. The first wave are those injured by falling debris. The second are those injured in the cleanup phase. The last, of course, are survivors extracted from fallen rubble. People trapped in rubble are susceptible to kidney failure, for a slightly complex reason. When you're struck and pinioned by hard surfaces, the constant pressure on your damaged limbs causes the muscles to break down. Large protein molecules course through your body and clog your kidneys, which in turn cannot properly filter toxins from your body. Dehydration exacerbates this condition, known as *rhabdomyolysis*. That's why people who are rescued from rubble die a few days later.

But don't panic if you live on or near a fault line. Here are survival tips to keep in mind:

- During a quake, you might think that the wisest course is to flee a home, business, or hotel entirely and get in the open air where nothing can fall on top of you. That's easier said than done. Moving from one place to another exposes you to more peril because you'll potentially be dodging one obstacle after another. Also, the ground itself will be moving under your feet, making it difficult to walk or run safely. Stay where you are and look for the best shelter you can under the circumstances. Get under a sturdy dining room table for protection. In an office place, get under a conference table or desk.

- After the initial quake has passed, assess your surroundings. If you're inside a building and it looks undamaged and safe, stay indoors. If the structure is damaged, leave. If you're outdoors and a structure appears damaged from the outside, don't enter it for any reason, even to retrieve that one little valuable. Aftershocks are common, and sometimes cause greater loss of life because people have begun cleanup and rescue, and are working or moving around in the vicinity of unstable structures.

- If your home was damaged in a quake, you may be tempted to inspect or shut off the water, electrical service, or natural gas valve. Resist that urge until you know for certain that the structure is safe.

- If you're in a coastal area with tsunami potential, leave the area immediately and seek higher ground.

- If you are trapped in building debris, do not light a match or use a lighter to inspect your surroundings. Natural gas leaks are common. Do not scream for help. You'll only exhaust your energy and inhale surrounding dust. Look for sustainable ways to make noise, such as using a whistle if you have one, or using your keys or mobile phone to tap nearby objects. Tap out SOS in Morse code, which is the international distress signal. The signal (* * * — — — * * *) consists of three short taps, three long taps, three short taps. (To familiarize yourself with how this might sound in real life, look for audio examples online to listen to. Also see page 48 for the full alphabet.)

- If your phone is capable of making calls, text or phone for help, but be aware that you may not be able to describe your location to emergency personnel with any degree of precision. Your best bet is conserving your strength, sipping water slowly if you have it, and focusing on making regular, sustained noises.

FLOODS

It's an unfortunate fact of our changing world that more humans are living in flood zones than ever before. Sea levels are rising. Rain events are becoming more extreme. In the United States alone, floods are now the most common and most disastrous event that can befall a region. Billions in property damage are caused each year by floods.

The city of Venice, Italy, famous for its canals and bridges, is no stranger to water. High tides have flooded the city for thousands of years. It's a common sight to see knee-deep "acqua alta"—high water—covering that famous piazza in front of St. Mark's Basilica. For centuries, shopkeepers have marked lines on their buildings to keep track of how high the tide reaches. In 2019 the water hit the highest marks anyone had seen since 1966. A floodgate project, designed to keep back the sea, has been in the works seemingly forever. It's only recently that that public works project has reached the point where they've been able to test the gates to see if they work. They do, but the work is still ongoing. Time will tell if that city, and many others that are threatened by rising sea levels, will survive.

It used to be that you were more at risk if you lived on a coast, but that easy assumption has been challenged in recent years, as more *inland* locations are threatened by rising water, and scientists and insurance companies are being forced to rethink flood zone maps. Even if you were told when you bought your home that you were not in a flood zone, it pays to investigate whether the situation has changed.

You'll also notice that not all nations in the world are prepared equally well for massive storms. Hurricane Catarina back in 2004 was an aberration in the sense that it was the only hurricane-strength storm on record to have formed in the South Atlantic. When the hurricane came ashore at Category 2 strength, the beautiful nation of Brazil was unaccustomed to storms of such intensity, and was not prepared the way nations in the Caribbean or the American South are. As a result, 36,000 homes were either destroyed or damaged, and the region lost the lion's share of its banana crop that year.

Floods result from heavy storms or post-hurricane runoff. The obvious survival option in a flood is to get to high ground as quickly as possible. The people who wait to evacuate are the ones we later find stranded on rooftops, in trees, or in disabled vehicles.

The worst thing you can do is wade into standing water. It's not as easy as walking across the low end of a swimming pool. Floodwaters rage out of control, with powerful currents ripping through them—a macrocosm of a tiny stream or creek. If you must walk through floodwaters, use something to brace yourself, like a walking stick. Even then, you face unique challenges that people often fail to expect. Water carries debris at torrential speeds. You could be hit with wood ripped from a home that is studded with nails, or get rammed with uprooted trees or road signs. (See the instructions on page 31 for treating impalements.) You could easily stumble on submerged objects and be swept away. The water is *always* deeper than you think. I was on the ground after both Hurricanes Katrina and Rita. What amazed and terrified me was seeing how massive mountains of water had uprooted coastal homes from their foundations and shoved them into a massive stacked pile of rubble, as if they were mere toys. Storm surges are the most dangerous part of the hurricane cycle.

- In general, the rule of thumb in hurricane conditions is, *"Run from water, hide from wind."* That means that if you're in a dry place, and you suddenly realize that water has been released, broken through floodgates, or surged on a coast near you, you must flee. This is based on the obvious fact that humans can't survive under water. There is no point in waiting to see how high the water "really" gets. Any sizable amount of water is usually a bad thing. Get out of there! In contrast, if all you're experiencing are high winds, it is smarter to take cover and shelter in place. High winds are more dangerous when you are out and about, because you can easily be struck by flying objects.

- If you attempt travel in floodwaters, keep your immersion brief. If you're swept under, try to turn your body so you are faceup, head above the water, and traveling downstream feet first. Keep your eyes open so you can see what's coming up ahead. This position—a kind of floating backstroke—will theoretically protect your head from colliding with obstacles. Most people have more strength in their legs than their arms, and you may be able to kick yourself away from oncoming obstacles. Notice I said *may*. Stay alert. Watch where the currents are strongest, and try to paddle yourself away from them and toward the "shore," or wherever the water appears less ferocious.

- Resist the temptation to wade in to help family members, pets, or neighbors. It is smarter to stay on high ground and throw the humans a rope, a stick, or some sort of object that can be used as a floatation device. Pets may not realize that you are trying to help them, and you are better off waiting until rescuers arrive in a boat or raft to approach your animal.

- Do not attempt to drive through standing water. Automobiles are not impregnable. Six inches (15 centimeters) will stall a car. Twelve inches (30 centimeters) will make a car float. Any higher, and the car can easily be swept away in the floodwaters. Don't remain in a floating vehicle; you have no idea where it will go, or how it will end up. There's absolutely no guarantee, for example, that its tires will remain down and the passenger cabin up. If the car starts to sink beneath the water, your best way out is through the windows. Doors are hard to open against the crush of water. Roll the windows down immediately, and crawl out. If you wait too long, it will be hard to get the windows down because water will be pushing against the glass.

If you have power locks and windows, the circuitry will fail when wet. Your only way out is to break the windows. Storing one or more rescue tools in your glove box and elsewhere in the vehicle is an excellent idea. These will allow you to sever seat belt restraints and break the car's tempered glass, allowing you to escape. Bear in mind that that windshields are the strongest glass in your vehicle. They are designed to protect vehicle occupants in the case of accidents. They are made of rugged *laminated* glass, not tempered. This means the only windows you'll be able to break easily with most rescue tools on the market are the rear and side windows.

If you don't have a rescue tool, you can break your side car windows using the headrest. Start by pulling the headrest out of the seat back. (You'll need to press the release button.) Each headrest has two strong metal prongs jutting out of its end. Insert one prong in the bottom corner of a window, right where the glass meets the weatherstripping. With the prong digging into this spot on the glass, lever the cushioned headrest down toward your lap. The glass will shatter, allowing you to escape.

All of this becomes far more complicated if you have family members, especially children or loved ones who may be unfit or elderly, in the vehicle with you. But not everyone will be able to get out without assistance. Don't make matters more difficult for yourself and the ones you love. Avoid deep water when you're driving a car.

- **The time to flood-prep your home is** *before* **the storm occurs, not during. You do not want to wade into a darkened basement that has been filled with rainwater in order to, say, prop up your nice couch or gaming chair on cinderblocks. If the structure has not lost power, you could easily be electrocuted. At the very least, you could lose your footing and hurt yourself. It's very difficult for untrained family members to fish out a loved one in the dark. Don't put yourself or others in that position.**

- **Once you're out of danger and back on dry land, don't remain in your wet clothes. During a flood, the water becomes contaminated with every liquid in the city or neighborhood. Gasoline. Antifreeze. Sometimes sewage. Shower, dry off, and wash or dispose of those contaminated clothes. Though humans love water and take to it easily in recreational settings, our skin is not designed to remain wet for long. Prolonged contact can cause rashes and tissue damage. Get dry and get checked out.**

MUDSLIDES

At first thought, you'd think mudslides go hand-in-hand with floods, major storms, and hurricanes. And you'd be right. But there are far more reasons why the earth suddenly gives away. Human-made reasons *and* natural reasons. Wildfire, for instance, denudes the landscape, leaving the ground unprotected and vulnerable to rainwater. That's a natural cause.

In 2019, 12 million cubic meters of mud were released when a dam belonging to an iron ore mining concern in Brazil failed. This firm had already been under investigation for another dam disaster—Brazil's worst environmental disaster—back in 2015. The second mudslide killed about 270 people, and the spill released untold metals and mining debris into the surrounding water basin. Those two cases were both related to the structural integrity of dams, and it was easy to point a finger at the negligence that caused them.

But most mudslides have less egregious causes. You can't blame just one person or entity. The blame must be laid at the feet of humans in general, and the seemingly endless march of urban sprawl. Humans building homes, businesses, and new developments on slopes and exotic locations remove the vegetation that is critical to keeping the ground stable. Along comes a major storm and—whoosh!—suddenly an avalanche of mud and dirt washes down from higher ground. The debris field carried by a mudslide can be horrifying. Homes, vehicles, chunks of road bed, uprooted trees, and dislodged boulders could all be part of the terrible matrix.

- **If you're living in or visiting a mountainous or hilly location where abundant rainfall has occurred, keep an eye out for the danger signs. Streams will be swollen with brown, churning water. Trees that were upright days ago will be leaning. You might find pools of water collecting in unusual locations. Slopes or lawns will have cracks in them where the ground has shifted very slightly. The foundations of homes and business structures may have developed cracks, or become more exposed. Your home or the place where you're staying** might make groaning sounds, indicating the ground is settling. These are all signs that you ought to tune in to the news and see if any mudslide warnings have been issued. It may be wise to leave the area immediately, without waiting for an evacuation order.

- Strap on a bike helmet or football helmet if you have one. It's probably the best thing you can do to protect your head for what's coming. It might look goofy, but at least you can laugh about it afterward.

- If you do flee in a vehicle, stay alert to what's happening with the road and the ground around you. If a slide happens, and strikes the vehicle, you have to decide whether to exit the vehicle or stay with it. Only exit if you feel it is safe to do so. Ideally, you want to be able to get out of the danger zone on foot. If you don't think that's possible, stay in the car, keep your windows up, and get low so you are out of the way of objects that might penetrate the windows or windshield. If the vehicle is still operational and upright, you still should not attempt to drive through a thick layer of mud. It's as damaging to the vehicles as driving through floodwaters.

- If you are on foot, and perceive a slide heading your way, don't run straight away from it. Mudslides are made up of sodden mud and debris, so they are very fluid. You're not going to be able to outrun it. Instead, do this: If the slide is moving north, you want to run east or west, along the edge of the flow.

You want to get to the spot where the flow of mud tapers off and the ground is holding. Yes, I know this is more easily said than done. You have no idea how far you'll have to run to get to the safe zone. And you might well have no idea which direction—in my example, east or west—is the safest direction to run. It's a total judgment call. Higher ground is better than lower ground, obviously. But whichever direction you go, you will most likely be traveling on *uneven* ground. Prepare for it. Adrenalin's on your side. Think quickly and act.

- If you are hit by the mud or are about to be hit, curl up and protect your head at all costs. (You have your helmet on, right?) Try to keep an air pocket above your head, pushing the dirt and debris out of the way with your arms. Resist the urge to scream! Mud will rush into your mouth and suffocate you. Once they come to a stop, mudslides dry quickly. If you haven't already, you want to fight to get yourself some breathing space before the mud surrounding you becomes stiffer, drier, and more unmanageable.

Next, assess your body's orientation. If you're unable to see daylight, you may not know which direction is up. To figure that out, try this. Allow a little spit to drool out of your mouth, and watch how it behaves. If you can't see it, feel for it. Where the drool drops is *down*. You want to dig *up*. If you're not able to get completely free of the mud, or you are caught on debris, call for help when you emerge from the muck. Conserve your energy and wait for help.

I would not want to be in a home or other structure when a mudslide hits. The ideal is to evacuate as soon as you are ordered, or sense there is potential for a disaster. But if you are trapped in a home, the best place to be is on an upper floor, so you're theoretically away from the strike zone—the home's foundation—when the slide hits. (Understand that in a slide, the top of the house could easily become the bottom.) The best-case scenario: The house rocks but stays put, and the mudslide moves around it or occurs on the slope below it. Get to the interior of the structure, away from windows or upper-level doors that open onto balconies, decks, terraces. Try to stay away from heavy objects or furniture that can shift in case the house slides off its foundation. Seeking shelter in an immovable object like a bathtub might be a good idea. From here, take all the same steps we discussed for outdoor survival. Curl into a ball, protect your head, focus, and be prepared to build that bubble of air around yourself if you need to.

If you ever come upon a mudslide victim who is not responsive, begin CPR immediately. Yes, it's entirely possible that their nose and mouth have become impacted by the mud, but I would not sweep the mouth or probe for blockages unless that mud were completely visible, and obviously causing problems. If the person is suffering impalements or other wounds, have another bystander or Good Samaritan treat those while you attempt resuscitation. Stay with the person until professional help arrives so you can bring their needs to the attention of medics, or swap out duties.

MUDSLIDE OR FLOOD? THE GREAT MOLASSES DISASTER

In January 1919, a massive steel tank on the grounds of a distillery burst in Boston, spilling more than 2 million gallons of molasses onto city streets. A great dark wave rolled in every direction, tearing houses from their foundations, flinging streetcars, smashing girders holding up elevated trains, and drowning humans, pets, and horses. In some spots, the depth of the liquid rose to 15 to 25 feet (4.5 to 7.5 meters). The low end was 2 to 3 feet (0.5 to 1 meter). Innocent bystanders never had a chance, as the torrent washed over them or carried them away. More than 21 people died; more than 150 people were injured. The strange disaster took nearly a century to investigate and properly understand. The distillery had recently received a large shipment of molasses, intending to refine it into alcohol for a variety of industrial and commercial purposes. The tank they stored the molasses in had never been filled to such a capacity before. Rivets holding the tank together failed, unleashing a 35-mile-per-hour (56-kilometer-per-hour) tsunami of goo. Modern scientists observe that molasses is 40 percent denser than water, so the released liquid behaved more like a mudslide than a flood and delivered an unprecedented level of destruction few people could have seen coming. This was the ultimate freak accident, an example of how none of us can ever say we've seen it all. There's always something weirder around the corner.

BLIZZARDS

Back in the day, my Air Force Academy buddies and I had just finished our exams and were so anxious to get home that none of us paid much attention to the news. Not that we would have, maybe. We were young and thought ourselves invincible. On a darkened road in the middle of New Mexico's wintry vastness, the black sky turned white in seconds. The snow blew fiercely, erasing visibility. But we were Air Force cadets. We thought we could get out of any hairy situation. So we gunned the engine, determined to get out of there before it got worse, forgetting that our Chevy Malibu was not equipped with radar like the planes we would one day fly.

One of our tires hit an icy patch. In a flash, the world spun around us in a 360-degree blur, and when we came out of it, the car was in a ditch.

In the New Mexico desert, you sometimes find people who like their solitude. We found one of those guys at the first door we knocked on. He chased us away from his house by sticking a shotgun out the window.

The night ended with a state trooper kindly calling a tow truck for us. As we shivered by the side of the road, my buddies volunteered me to talk to the tow truck operator because they had the notion that I was the single best negotiator of the bunch of us.

"How much is this gonna cost?" I asked the dude as he was pulling out his chains.

"How much you got on you?"

I checked my wallet, and asked my friends to check theirs. "Forty-four bucks," I said, "and that's everything we've got, mister."

"Fee'll be $44."

Turns out, I'm the world's worst negotiator.

The human animal is designed to live through extremes. We conquered this planet because we are naturally adaptable to hot and cold climes. But without shelter in a blizzard, things will go south quickly. You don't have to be in the Arctic Circle to perish or get frostbite. Look at us: We were in the middle of New Mexico, a place that most people associate with downright warm temperatures. And it is—in summer. Remember the Rule of Three? In winter, the high desert will kill you in 3 hours if you can't cover up or find shelter.

In the best of all possible worlds, you'll see a blizzard coming. Weather announcers will be going on about it for days, and you'll have the heads-up that you need to lay in all the provisions—food, fuel, snow shovels, road salt, and other items—you need to keep your family safe.

But there are several twists to this story. If you live in a part of the world where snowy weather is common, you've probably seen countless scenarios where the predicted storm turns out not to be as bad as the experts said it would. (That's not great because it trains people to ignore forecasts.) Equally common is the teensy storm that grows beyond expectations, leaving people and snowplows digging for 3 days later. And there is the issue of *relative* snowfall. A Wisconsinite scoffs at 3 inches (7.5 centimeters) of snow; Georgians not so much.

From the doctor's perspective, the critical component of a blizzard is not the amount of snow, per se, but the amount of wind. Gusts demolish visibility, increasing the chance that you'll end up in an emergency room. Harsh wind punishes human skin, upping the wind chill factor and the chance of frostbite.

Not matter where you live, no matter where you travel, you must take blizzards seriously.

Surviving a Blizzard Even If You Have Shelter

A warm home is the safest place to be in a blizzard. The biggest threat in this case is loss of power. You may have a working fireplace or wood-burning stove. You may have a backup generator to run critical appliances.

If you have none of these things, you still have the advantage of being sheltered and dry. You have access to dry blankets and multiple layers of clothing. And you have the company of friends, family members, and pets—all of whom can and will assist each other in staying warm by virtue of their body heat. Even if you must pull on five pants and six shirts to stay warm, you are safer indoors than out.

Some things to look out for.

- The middle of a raging blizzard is not the time to perform all the essential tasks you should have done before the storm started. Stock up on food and other essentials as soon as the storm is predicted. Can anything on your property be damaged by heavy snow and wind? Button it up now.

Consider how you will attend to the needs of any pets or livestock in your care. Protect sensitive plumbing from the possibility of freezing. Disconnect all outdoor hoses as well.

- If you have a backup generator, check to see that it's operating correctly and you know how to run it. The best type are fixed, outdoor systems tied to a home's natural gas line and directly wired to your home's electric utility box. In case of an emergency, they'll switch on as soon as the power goes out. Generators that run off gasoline can and should only be operated outdoors or in a structure that is not attached to a home. The exhaust from these devices can kill. Make sure you have all the peripherals you need to tie the device to your home's essential circuits without bringing the generator into an indoor space.

- Familiarize yourself with the signs of hypothermia and frostbite. Be ready to assist family members in trouble. Every year a certain number people die from hypothermia in winter, even though they were indoors. If a home is not insulated properly, or if a loss of power is protracted, room temperatures can drop to life-threatening levels. Be ready to move the entire family to a single room with blankets, water, and food. If you have a source of heat, such as a fireplace or wood-burning stove, so much the better. Hunkering down in one place is the best way to keep an eye on everyone at the same time. Hopefully, your confinement in the dark will be brief.

- People who have natural gas stoves think they can still cook if they lose electricity. But in fact, gas stove igniters—the tech that creates that *click click click* noise prior to sparking a flame—actually run on electricity. You will need a good supply of *long* matches or a long-nose lighter to be able to cook or boil water. If you go this route, be safe about it. Never run a natural gas stovetop or oven for the sole purpose of heating. I don't care how bad it gets; this is far too dangerous to be considered a survival option. Prepare food to feed the troops, then shut the burners off.

- Keep flashlights or battery-powered lanterns handy. Avoid candles. Conserve mobile phone batteries. (Texting uses less power than phoning.) Keep your crank radio charged and ready to listen for important news.

- What to listen for: If roads are cleared but power is not restored, authorities in your city or town may arrange for locations where you and your family can shelter to stay warm. That's the only reason you should consider leaving your home. If you leave your residence for a warm local shelter, make sure you are not leaving on any appliances or utilities that could cause problems if power is restored in your absence. Gasoline generators should never run while everyone's out of the home.

Surviving a Blizzard in Your Car

Every vehicle you drive in wintry weather must be equipped with essential gear to help you survive if you become stuck on the road.

- ☐ A fair amount of fresh water for you and each person traveling with you
- ☐ A little food
- ☐ A good emergency blanket, the kind that folds up small but shrugs off cold
- ☐ Extra layers of clothing
- ☐ A mobile phone charger
- ☐ Snow boots
- ☐ A flashlight or headlamp, with fresh batteries or freshly charged
- ☐ Jumper cables with portable battery
- ☐ Folding shovel
- ☐ Disposable hand warmers
- ☐ Some sand or cat litter to give your tires some traction to pull out if you get stuck

Everything on this list can be safely stored in your vehicle during the winter months except water, which will freeze. Fresh food faces the same danger, but some non-perishable protein bars or ready-to-eat meals will be fine if kept in the heated cabin with you. Since there's an argument for taking food and water with you in any season, especially on long trips, one option could be to keep your stash of food and water in a compact backpack that you bring in and out of your home. Eat or discard them as the seasons turn, and replace them with new stock.

Understand that four-wheel drive does not mean "four-wheel stop." Just because you can drive fast on ice doesn't mean you will stop fast. In a storm, an SUV is just another big vehicle looking to collide with whatever's out there. If you're not accustomed to driving in snowy weather, seek training or minimize your travel during certain types of weather.

If you are stuck in a storm and cannot move your vehicle, run your engine just enough to warm the cabin, then shut your ignition to conserve fuel. (Ten minutes is usually sufficient.) Do this for as long as you need to. Use your emergency blanket to extend how long you can wait to warm up again. Never do this if your exhaust pipe is covered with snow. You really ought to be checking it on a regular basis to make sure snow hasn't fallen from, say, a nearby snowbank, and blocked it. Carbon monoxide will flow back into the cabin and kill you.

Do not leave the vehicle to walk for help unless you can see your destination from the windows of your car, and know for certain that help is likely. You don't want to risk walking even a short distance in the cold only to find that the commercial business you headed to is locked up tight for the night, or the home you've walked to is empty. Remember that the authorities are more likely to find the car before they find you, and they will expect you to be inside the vehicle.

Even in a cold car, you need to stay hydrated. You can drink snow provided you have let it melt first. Don't eat it while it is still in its frozen state. It will drop your body temperature too precipitously.

In cold weather, some clothing works better than others. A wool sweater, for example, will still keep you warm if it has gotten wet. So will fleece. Some performance shirts are designed to wick away water from your skin. Wet cotton kills. It will literally freeze around you. Switch to something warmer if such a shirt becomes wet from snow or sweat.

- If your fingers feel numb or tingling, tuck them under your armpits and keep them there.

- Body heat can keep you alive. If you're sharing the vehicle with someone else, get under that blanket together. Not time to be bashful.

Understanding Hypothermia and Frostbite

The human body faces two serious dangers in cold environments: hypothermia and frostbite.

Hypothermia is when your body cannot produce enough heat (or energy) to maintain your ideal human body temperature, which is around 98.6 degrees Fahrenheit (37 degrees Celsius). The colder the surrounding air is, the harder your body has to work to maintain that high core temperature.

Even under duress, the body is a complex machine, and it resorts to an almost ruthless protocol to remain alive. "Okay," it says, "there's no way we can stay alive in this cold. The heart and brain are the most important organs. Anything else can be sacrificed."

So what does it do? It starts shutting down the parts of the body that are farthest from the heart. Those are the extremities; doctors are taught to remember them with this little rhyme: "Fingers, toes, penis, nose." In a healthy body, blood routinely flows to the very tip of these parts, and then doubles back to be refreshed with oxygen by the heart.

When hypothermia looms, the body is not going to waste time pumping blood all the way to those tiny tips. It conserves energy by keeping blood flowing through the torso. The result: frostbite. In the early stages of frostbite, the sufferer feels numbing and tingling in the extremities. As frostbite becomes more severe, the extremities are beginning to freeze. They become immobile and take on a waxy, wooden appearance.

Body temp begins to drop, marching through the three stages of hypothermia.

- The first stage, *mild* hypothermia, is marked by shivering—which is a good thing. The muscles in your body tremble in an attempt to warm up your body. Believe it or not, at this stage, your body temperature is within 90 to 95 degrees Fahrenheit (32 to 35 degrees Celsius). I know! That sounds pretty *warm,* only a few degrees from the optimal level, but this shows you how such a small drop will trigger your body to protect itself.

At the second stage, *moderate* hypothermia, the body's movements become slow and uncoordinated. The sufferer may appear confused, and their speech slurred. Internal body temperature is between 82.4 degrees and 90 degrees Fahrenheit (28 to 32 degrees Celsius). At this point, the person is in grave danger because they're so impaired that they may be unable to take decisive physical action to change their situation. For example, they'd find it really hard to build a fire, which is a fairly complex task. If they are outdoors, they may not have the mental bandwidth, not to mention the physical stamina, to stagger in the right direction of protective shelter. For complex physiological reasons, the person feels warm even though they are not, and may begin to undress. "I've got to get out of these clothes," they think. "I'm burning up!" Not true at all. Here's what's actually happening: The body's ability to thermoregulate itself is failing rapidly. All the blood that rushed into the body's core to protect the heart and brain is released in a huge rush, and is pushed outward to the surface of the skin. That's the reason the person feels warm. But the truth is, if they're alone, without anyone to assist them, it's bad news.

As soon as their internal body temperature drops below 82.4 degrees Fahrenheit (28 degrees Celsius), they've hit the third stage, *severe* hypothermia. The person's heart fails and they slip into cardiac arrest. If they don't receive CPR, they'll be dead in 3 to 5 minutes.

Surviving Hypothermia

When we get hypothermia victims in the ER, our treatment is focused on warming them at every possible level. We cut off their clothing to minimize hurting their fragile skin and pack them in a Bair Hugger gown or blanket, which circulates warm air around them. If we have to use a ventilator, we pump slightly warmed air into their lungs. All intravenous liquids are slightly warmed as we pump them into the victim's body.

If you're in a remote location and unable to get the person to a doctor, you can easily do the following.

- Get the victim into a warm space. You don't need a raging fire. A normal room-temperature living space is better than the cold they experienced outdoors.

- Get the person to undress. Cold or wet clothing is only making things worse.

- Cover the person with a blanket. If you have a few others and can spare them, use as many as you can to drape them and keep them warm. Mittens are good too; they'll keep fingers warmer than gloves.

- Remember: Hypothermia causes confusion. The person may resist all these efforts to help them. They might not comprehend what you're doing, or why. They may insist that they're fine, and too hot to need all these blankets. Be prepared to be forceful.

- Do not permit food or water to pass their lips until they are able to feed themselves, or lift a cup or glass to their lips. I know that it feels wise to help the person drink hot chocolate or warm soup—but just don't. You can induce vomiting, choking, or burns if you try to feed someone food when they're not aware enough to judge for themselves if the food is too hot or too much. Better to have those things ready when the person asks for them.

- If you're alone, you must take these steps without assistance, even if your body is fighting you at every turn. If you can get to room-temperature shelter and huddle under a blanket, you will survive and be able to fend for yourself once your core temperature rises.

Surviving Frostbite

Frozen extremities are like meat in your freezer. The more you defrost and refreeze them, the greater the chance that the delicate cellular walls will rupture.

For that reason, frostbite complicates our survival plan. If you are in transit from an outdoor location or a cold indoor location, and there is a possibility that the victim's fingers, toes, and other extremities will refreeze, do not attempt to treat their frostbite. Just protect their limbs from getting colder.

You really must wait until you have reached safety, and are secure that the danger has passed. If you have fled your cold vehicle and taken shelter in a drafty, unheated home or barn on the roadside, you are not out of danger and should not attempt to treat frostbite.

Here's what to do:

- **Do not rub snow on frostbitten limbs. This is a dangerous myth that will only make matters worse.**

- **Treat the person for hypothermia first, then treat the frostbite. It makes no sense to save the fingers of a person who is about to lapse into cardiac arrest.**

- **Prepare a warm water "bath" for the victim's fingers or toes. Fill a bowl with warm water and test the temperature as you would for an infant's bath. Touch the water with your elbow and see how it feels. It should feel very warm, but not scalding hot.**

- **Dip the person's extremities in the warm water. Keep them in the warm bath. Change the water when it gets cold. If the person is suffering frostbite on fingers, toes, and other extremities, you can treat one set of digits while another person helps with the others. If you're the only person around, pick your battles. Treat the extremities that look worse, and then switch out.**

- **Cease the warm water bath when the fingers, toes, or other extremities begin to change color. This means that they are warming up.**

- The victim is not out of danger yet. Their body is waking up, and discovering the extent of the damaged tissue. Blisters and pain are a good thing, just as they are with burns. (See pages 25–26.) It means the underlying nerves have not been destroyed, and only the top layers of skin have been damaged. Cover the blisters with loose bandages and administer whatever over-the-counter pain relievers you have. No pain may mean that the situation is worse. Think about how soon you can seek professional medical attention. Pain or not, any incident of frostbite should be carefully inspected by a doctor.

AVALANCHES

Before we talk about avalanches, let's talk ice cream. I want you to picture a cone of soft-serve upon which has been drizzled a particular type of chocolate sauce. It's the kind of sauce that quickly dries into a thin, almost crunchy shell of chocolate. The second you bite into the cone, what happens? The chocolate sheet cracks and slides off the wetter, melting ice cream underneath.

Avalanches can be like that cone. There's one stiff, cohesive sheet of snow and ice, layered on top of a weaker, wetter layer of snow. All it takes is a sudden increase in weight. A fracture occurs, and the whole structure gives way. The top layer slides down over the weaker layer in one terrifying piece, and doesn't stop until the slope of the mountain levels out and the torrent of shattered snow runs out of steam.

That's just one type of avalanche. The most common type.

Sometimes the snow isn't in a giant slab. It's a loose pile that was deposited in a recent storm that never set up properly. A little warm weather, and the snow melts a little. Along come a handful of skiers, snowboarders, or snowmobilers. The snow can't take the extra weight, so it gives way.

Ski areas and winter tourist spots have gotten very good at mitigating avalanches. They detonate the wild areas, hoping to shake off the most fragile sheets of snow. They cordon off the "back bowls" on their property, forcing skiers to stick to areas that are stable or have been built with man-made snow.

Next time you're in a ski area, read the back of your lift ticket. It's right there in black and white. When you hit the slopes, you're acknowledging that you understand you're taking your life in your hands. Earth is a giant orb that is constantly changing. Dynamic processes are always at work. There's no guarantee the snow beneath your feet will stay put.

If you're stuck in an avalanche, you stand a good chance of being killed by blunt-force trauma. After all, you're being pelted by boulder-size ice, snow, rocks, and the debris of fallen trees that's flinging your very fragile human body down a mountain at speeds of up to 80 miles (129 kilometers) per hour. If you survive that, you're incredibly lucky. But you've still got a long way to go before you're sucking down that mug of hot cocoa and marshmallows back at the lodge.

If you're trapped alive in an avalanche, you don't die right away. First, the weight of the snow immobilizes you. It sets very quickly, hardening to a consistency of concrete, or, well, frozen ice. There you are—trapped, unable to move, your heart racing, your breath puffing away into this snowy cocoon. With every breath you take, you deplete the oxygen in your tight pocket of underground air. You suffocate very, very slowly on your own carbon dioxide.

It's a terrible way to go. But if you have at least thought about the possibility of avalanches, there's a slim chance you could have avoided this disaster entirely. Here's how.

- Always stay alert on snowy slopes. Remember my maxim: Keep your head on a swivel! If you're not there to ski, I fail to understand why you need to be there in the first place. Snowy, uphill hikes aren't fun. And snowshoe treks or cross-country skiing are best conducted on level land, where avalanche risk is largely eliminated.

- When downhill skiing, stick to the designated slopes. They're there for a reason.

- Avalanches are triggered by a sudden increase in weight, but you'll have little to no idea how much is too much. If the ground below your feet fractures or starts to give way, you can hop above the fracture line in an effort to be above the zone that's about to plunge down the slope.

- If the snow is above you and already sliding down, you're unlikely to outrun a torrent of ice that's moving faster than a car on a highway. Propel yourself so you're skiing perpendicular to the debris. (This is similar to the advice on surviving mudslides. See pages 111–114.) In other words, if the avalanche is moving south, ski in an easterly or westerly direction.

- If you get caught in the debris field, release and kick off your skis. Start "swimming" instead. Use your arms to paddle and propel yourself ahead of and to the top of the rushing snow. Keep your head up. You want to keep daylight in your sights at all times. If you don't fight, you'll be tossed like a rag doll and the debris will simply fold over you.

- Eventually the avalanche will slow down. But when it does, it will begin to harden quickly. Seconds count. If you are not on the top of the snow, scrape to form as big an air pocket as possible in front of your face. Chances are you will have lost your hat, gloves, and possibly shoes or boots. To protect your hands as you dig, pull them up into your jacket sleeves. For this reason, it's not unreasonable to carry a second pair of gloves in an inside pocket of your jacket if you know you're venturing into an area where avalanches are a possibility.

- If the snow has not hardened—and that's a big if—dig yourself out. Do all you can to get an arm sticking out above the snow. That's how your rescuers will be able to find you quickly.

- If you don't know which way is up, use the trick we described in the mudslides section. (See page 113.) Let some spit drool out of your mouth and see which direction it drips. It will drip down. You dig up.

- Keep your breath calm and even, so you consume oxygen slowly. Try to minimize skin-to-snow contact to reduce the chances of frostbite. Resist the instinct to yell, lest your mouth fill with snow.

TORNADOES

I magine an airport runway stretching off into the distance, as far as your eye can go before it meets the rising sun. Now, imagine that runway covered with the furry bodies of hundreds of rabbits.

Dead rabbits.

That was the sight that greeted me the morning after I was caught in a tornado. My wife and I had survived, thanks to some quick thinking. But the world felt completely changed when I arrived at the Air Force base the next morning. The tornado that whipped through our Texas city had spared the base's airplane hangars, but had somehow ripped up an underground warren, resulting in spewed rabbits everywhere.

That's how tornadoes are. You never know what they're going to do. In a flood, as we saw, the danger lurks underwater. In a tornado, the danger is airborne. You can be easily impaled or broadsided by an object large enough to whip a car off the road.

Everything around you has the potential to turn into a projectile that can maim and kill. Homes are wrenched from their foundations or pounded flat. Trees are ripped from their roots. Boats stored on trailers are sent careening down highways. Giant pieces of structures peel off and become flying, nail-studded weapons wielded by the unseen hand of Mother Nature.

I've heard stories of farmers on the Great Plains who look out the windows of their homes and watch for tornadoes with their phones in one hand. Why? Tornado insurance is expensive, but insurance firms will cover you in minutes if you call before one hits. No one likes to spend money if they don't have to, so the farmers adopt this bizarre, wait-and-see, phone-in-hand vigil.

That's about the only lucky break you get with tornadoes. Authorities will warn you when conditions are likely to generate these harrowing twisters, and you can react accordingly.

That's what happened to me and my wife.

The year was 1982. The city, Lubbock, Texas, where I'd gone to continue my training as a fledgling pilot in the U.S. Air Force.

Lynda and I had met the year before at a party in Monument, Colorado. She was a liberal college student studying at the crunchiest, most granola-eating city in Colorado: Boulder. And I was the conservative Air Force cadet, looking trim and fit in his uniform and buzzcut, training at the academy in nearby Colorado Springs.

On paper maybe it didn't look like our romance would work, but we had a lot in common. For one, we were both Latinos, and had grown up speaking Spanish with our parents and grandparents. Lynda was an Army brat; I was an Air Force brat who'd been born in Greece, where my American dad was stationed. My parents were both from New Mexico, a long line of proud people who traced their heritage back centuries. They raised me and my three sisters in various cities in Europe, where Dad was stationed. I was 16 years old when I returned to my native country, the United States.

All I ever wanted to do was fly planes. As far as I could see, it was a proud job, possibly the world's coolest.

Air Force cadets are not permitted to marry, so Lynda and I waited until I graduated with my degree in engineering mechanics. In a sense, I had to earn my wings before I could put a ring on my bride's finger.

I had to do a year of training in Lubbock, flying T-37s and T-38s, fighters/trainers that are great fun to fly.

The night before the rabbits, we had some friends over for dinner at the condo we were renting. It was a stormy weekend night. Music blared from a transistor radio. Drinks, snacks, the whole bit. On weekends when we weren't working, pilots and their families partied hard.

And then, without warning, the radio started screaming.

Alarm! Alarm! Alarm!

As soon as I heard that ear-splitting screech, I knew exactly what it was. There's only one thing that strikes fear in the heart of a Texan—*a twister*.

My Air Force survival training kicked in like an engine roaring to life.

"Grab some blankets! Everybody in a closet!"

In an ideal world, you'll wrap yourself in something that will protect you from being impaled. But that night in Lubbock, I couldn't think of a single thing we had in that apartment that could stop a hurtling piece of wood or steel. All I could think of was blankets and a closet door. We used what we had. We and our guests huddled in a pair of closets and waited for nature to take its course.

People tell you that a tornado sounds like a freight train chugging by your outside walls. I don't recall that at all. I just remember hearing a massive amount of wind that wouldn't let up.

Every second under those blankets, I thought—is this it? Will the walls hold? Will the roof collapse? Are we done for?

And then, after thirty excruciating minutes, just like that, the wind subsided. Our apartment complex had missed being devastated, but others were not so lucky. That's the freakish thing about tornadoes. They destroy and kill with ruthless precision. They'll cut a swath through a neighborhood, obliterate ten blocks, yet leave six others completely unscathed. City-dwellers often wave off the dangers of twisters, reasoning that their chosen hometown protects them from this peril. But extreme weather is bringing unusual events to new locations every year. No so long ago, a twister flattened a block in New York's Queens borough, leaving streets on either side unharmed. You must be prepared, even if you live in an area historically unaffected by twisters.

The morning after our brush with a twister, I had to fly out early. As I drove to work, I saw hundreds of telephone poles snapped like twigs, down to the ground. At the airfield, planes had broken loose of their tethers and were smacked up against each other like amusement park bumper cars.

And then there were the rabbits.

As I climbed into my cockpit, I saw a bunch of fellow pilots heading out to the tarmac with shovels in hand. Every pilot who wasn't flying that morning was pressed into the service of cleaning this post-apocalyptic carnage—picking up one dead rabbit after another so we could have an unimpeded use of the runway.

They have a saying in the Air Force. Admittedly it's a joke, but it carries with it the weight of truth. We call it "the pucker factor," and it refers to that indelicate part of the human anatomy that shrivels when fear takes hold—the sphincter.

"We'll, *that* was a hairy situation," one pilot might say to another when they're back on the ground. "How much seat cushion were you sucking up?"

That day in the closet in Lubbock, there were no cushions to be found, but I promise you that I'd have been sucking up 50 percent of whatever was there.

Some things you need to know about tornadoes:

- Take a moment right now to make sure that your mobile phone is configured to receive wireless emergency weather alerts. This is the best way to stay informed if and when a tornado warning is issued. People are not always checking the news while at home or in the office, but most of us these days are glued to our phones. If you get one of those alerts, use your time wisely and be prepared to take action if the situation turns worse.

- Get as low as you can. In a home, you can move to a basement level to ride out a twister. Basements are usually the sturdiest part of the house, with concrete or cinderblock walls and thick steel beams holding up the central floor joists. Move away from any openings—such as windows or walk-out doors—that a projectile could penetrate. Get in the center of the room and hunker down. If you have your wits about you, or time to prepare, take down to the basement protective gear—bike helmets, football helmets, or construction hard hats all work—for everyone in your family (to protect your heads), water and a little food, a flashlight, your mobile phone, and your crank radio. Unfortunately, you might not be able to tell if the twister has spiraled out of the area or extinguished itself completely. Use your radio, phone neighbors, and cautiously peer out of basement windows.

- If you do not have a basement, do what we did in that condo back in 1982. Put as many walls as possible between you and the great outdoors. You want to get to the center of the house, and hope that the surrounding walls stop any projectiles before they reach you. For the same reason, it's best to cover yourself with something thick. The more layers, the better. If you are living in a manufactured home, you ought to move well before the storm to a structure better suited to ride out a tornado.

- If you're outdoors on foot, try to seek shelter. There's no way you can outrun or outdrive a tornado.

- If you're traveling for long periods in a car, you ought to check the local weather from time to time just to be aware of what's happening around you. If you cannot reach shelter before a tornado appears on the horizon, park your car and move away from it, to a location that is lower than the road. A road ditch or culvert will do fine. Lie face down and cover your head and neck with whatever you have handy—a blanket, jacket, etc.

- Never park your car under a bridge or overpass. You might think it will shelter you, but wind accelerates as it whips through barriers. Because of a phenomenon known as the Venturi effect, the concrete and steel pylons holding up the bridge enhance the effect of the tornado as the funnel is strained through it. The wind actually gets *stronger*.

- The only reason to remain in a car during a tornado is if you absolutely cannot see a spot that is lower than where the vehicle is parked. Even if the car does not become airborne, the windows may be penetrated by flying objects. Roll up the windows. Get as low as you can in the vehicle and cover yourself with any many layers as possible. Clothing, blankets, luggage, food coolers—anything that will stop a projectile from striking you.

- Blunt-force trauma and impalements are the most likely injuries sustained during a tornado. Follow my instructions for treating multiple bloody wounds (page 29) and impalements (page 31). In general, if the impalement is superficial, and you can see the end of the object sticking out of the victim, remove it, apply direct pressure, bandage or apply a tourniquet, and get them to the ER. Exceptions: Remove nothing from an eye or from a torso that might damage other organs.

LIGHTNING

It was the middle of the day and I was standing on the roof of the hospital where I worked in New Mexico, watching the rain fall. Watching it, hearing it, and smelling it. I've spent so much of my life in the desert or in semi-arid environments that the sight of dark, heavy rain clouds sliding across the sky never fails to move me.

But then, just around the time I was starting to think about ducking under an overhang to get out of the rain, I saw a flash of light off in the distance. A jagged, sizzling bolt lit up the sky, then disappeared. A few seconds later, I heard a thunderclap that was the loudest thing I've ever heard outside a war zone. The building shook under my feet.

It's time to get inside, I thought. Yes, everyone likes watching rain fall, but lightning is no joke.

The heavy steel door to the roof clanged shut behind me. I shook off some of the rain and started down the stairs.

Back in those days, before mobile phones were a thing, medical residents all wore what we called our Chewbacca belt. It was a massive utility belt studded with pagers linked to various units and departments in the hospital. When you came on duty, you accepted the Chewbacca belt from the fellow resident who was going off shift, and you passed it to the next doc when you went home to get some shuteye.

As I was heading down the stairs, the Chewbacca belt began screaming. The trauma pager and a couple of other pagers went off. That's how I knew I was needed in the emergency room. I grabbed the railing and started taking the steps two at a time.

By the time I got back to my post, the ambulances had already arrived, carrying our two latest patients. Both of them were unconscious.

Two men. Two separate incidents. The same deadly cause.

They had both been struck by lightning.

Lightning is a disturbance in the natural environment. Not unlike static electricity, too great a charge has built up in a cloud and it has to go somewhere. It craves release, and the only way it can achieve it is by taking the path of least resistance to the ground. Only when that charge hits the ground will the excess electricity dissipate.

Typically, lightning seeks the tallest objects in the landscape. Power lines. Power stations. Rooftops. Trees.

But sometimes lightning has a brain fart, and hits humans if they are unlucky enough to be in the vicinity. Our bodies also happen to be filled with salty water, which helps conduct the powerful electricity discharged by lightning.

My first patient that day had been mountain climbing when he was struck. The second was a 21-year-old soccer coach who'd been riding his bike at the time of the blast.

They both looked bad. What lightning does to the human body is horrific. In its zeal to reach the ground, the lightning had traveled completely through the mountain climber's body, blowing off his hiking boots—*and the toes inside them!*

Besides his missing toes, the climber's skin showed signs of "ferning." These are strange tattoo-like markings, which look like the lacy pattern of fern leaves or the image of a jagged lightning bolt. These marks usually cover the part of the body through which lightning has traveled. Ferning happens because as the electricity passes through a body, it zaps blood vessels and capillaries, causing them to leak into the surface of the skin. Under the climber's skin I could see the fine network of blood vessels that had popped and discharged into his flesh. Luckily, ferning is only cosmetic. It often heals in a few days if the patient survives. The bigger issue was that both men had gone into cardiac arrest. To save their lives, the first medics at each scene had to ignore the bleeding and missing toes to focus on getting the hearts of both men restarted.

Conservatively, they had each been struck with *30 million volts* of electricity. A blow that strong decimates the electrical system that keeps the human heart functioning.

What I saw that day never left me. I will always love the dry terrain of the American Southwest. And I still love watching the rain. But I will never love it enough to keep watching from the outdoors after I hear thunder. Life's too precious, and lightning too unpredictable.

What could my patients have done that day to protect themselves? What can you and your family do? Some lightning rules of thumb:

- Since light travels faster than sound, you will see lightning before you hear thunder. In general, the shorter the period of time between the two, the closer the lightning is to you.

- Here's the math, if you really want to know: Count how many seconds elapse between seeing lightning and hearing thunder. Then divide that number by five. Five seconds between the two means the lightning is 1 mile (1.6 kilometers) away. Thirty seconds means it is 6 miles (9.7 kilometers) away. I have heard advice that you should get indoors if you cannot count to 30 between lightning and thunder. Whatever. Here's the thing: The math is very cool, and I think every parent should teach their kids about it. It's certainly a fun fact about our amazing natural world. But as a doctor, I really don't think you should waste time listening and counting *outdoors*. Why take a chance? The best advice is still the old adage: "When thunder roars, go indoors." (You can still count from the front window and enjoy the storm's terrible beauty.)

- The safest structure is a permanent, enclosed building that is "grounded." That means that the building has plumbing and electrical wiring that leads from the structure right into the ground. So houses are safe. Businesses and stores are safe. Structures that have had special lightning-rod systems installed across their roofs, down their peaks, and right into the ground are very, very safe.

- Open-sided structures are not safe. A porch, for example, is not safe because it's wide-open to the elements. Metal or wood garden sheds are usually not safe unless they too are equipped with plumbing or wiring that leads to the ground. If you have outbuildings on your property to house livestock and other valuable equipment, consider installing a lightning-rod system to conduct an electrical strike properly to the ground.

- An automobile is safe as long as you're completely inside it, with the windows rolled up, and have a hard top. A convertible is not safe. Nor is a golf cart. Nor is riding in the bed of pickup truck. Nor is poking out of the sunroof of a friend's car, getting soaking wet and yelling "Woo-hoo!"

- If you're swimming, get out of the water. Water is just as big a conductor of electricity as your body is. If you're at an "attended" public pool, the lifeguard(s) will usually insist that you get out of the water and not return until 30 minutes after the last thunderclap has been heard.

- *Under no circumstances* should you stand under a tree or another tall structure during an electrical storm. Lightning has a tendency to "break off" once it strikes an object. For example, it will strike a tree, then branch off and hit whatever's nearby. The morning after a storm, it's not unusual in rural areas to find a lightning-damaged tree surrounded by a number of dead cows who wandered over to seek shelter at the wrong time. The lightning that killed them zapped down the tree and traveled across the ground in rain droplets to the bodies of those cows. Don't be that cow!

- For the same reason, you must never lie flat on the ground in lightning storm. The strike can travel across the ground, right to you.

- If you are like the two patients I described, exercising outdoors at the time of a lightning storm, try to get to a vehicle or grounded structure immediately. If you are camping, leave your tent and get to your vehicle to wait out the storm. If reaching safe structures or vehicles in the time you have is out of the question, seek the lowest ground possible away from trees. Do not sit or lie on the ground. Instead, crouch low on the balls of your feet. This way, you will have little contact with the ground and make a poor target for lightning.

- Even if you are inside a safe, grounded home during a storm, avoid taking a shower. Stay away from electronics, or unplug them entirely. And stay off corded telephones, i.e., landlines. (Mobile phones are safe to use indoors.) If the home were struck by lightning, the electricity would have to travel through plumbing and electrical lines on its way to the ground. You don't want to be anywhere in that potential path, least of all wet and naked in the shower.

- For the reasons outlined above, surge suppressors are critical to protecting your expensive electronic equipment in the home. These days you can also have a professional electrician install special breakers in your home's electrical box that will indicate if your home sustained a damaging strike or surge. You can also install "whole-house" surge protection systems.

- Lastly, if you come upon someone who has been struck by lightning, begin CPR immediately while someone else calls for help. You can certainly check first to see if they are conscious, but I assure you that if they have been struck, their heart will have stopped.

To be specific, the lightning will have de-synchronized the lower part of the heart. This is called *ventricular fibrillation*. If you could peek inside the lightning victim's chest at that moment, you would observe that their heart was quivering like a bag of worms, incapable of sending blood and oxygen to the brain or anywhere else in the body.

I implore you: If you have no training in this area, use the Bee Gees "Stayin' Alive" method I outline on page 33. Just pressing on their chest in the manner I showed you will "teach" the victim's heart how to start properly pumping again, and will potentially save their life. Don't waste time deciding if you should intercede. A lightning victim has 3 to 5 minutes to restart their heart, or risk death or permanent brain damage.

BLACKOUTS

A few years ago the TV network sent me to Houston, Texas, to cover the floods that resulted from Hurricane Harvey, a Category 4 storm that slammed into Texas and caused massive flooding, power outages, and widespread destruction.

It wasn't terribly easy to find a hotel that would accept me, my producer, and our film crew. But we lucked out and found a place. Which was great. We spent the day driving out, shooting, and getting some good stories that showed how citizens were coping.

Nevertheless, a bizarre scene was waiting for us when we got back to the hotel. To run the computers at the front desk, the hotel's staff had started a gasoline-powered generator, and when they found that they did not have a long enough power cord to reach the desk, they did what no one should ever do. They left the generator running in the foyer.

The thing was spewing exhaust everywhere. Staff and guests were all minutes from asphyxiating on carbon monoxide and fumes. And weirdly, everyone behaved as if this was something they just had to put up with temporarily because of the power outage. Everyone was going about their business, all the while ignoring the exhaust—not to mention the annoying engine noise—of this dangerous piece of equipment.

"Hey!" I told the guy at the desk. "You have to shut this off *right now* and get it outside. You're going to kill people!"

During the summer of 2020, when Hurricane Laura slammed into Texas and Louisiana, more people died from the improper use of gasoline-powered generators than from the actual hurricane itself. One family died because, even though they placed the generator outdoors, they left one of the doors of their home partially open. An open door was all it took to cause a tragedy that could have been avoided.

This is why I travel with a personal carbon monoxide detector. It cost me under $30, and it fits perfectly in my checked luggage. Usually, I trot it out when I've rented a room at a small, mom-and-pop homestay or B&B.

The hotel scene I described was probably the most flagrant carbon monoxide risk I've ever encountered—and it was in a large, well-known hotel chain.

But I should have seen that coming. When the lights go out, people start doing crazy things.

Most of us have experienced short-term blackouts. The power cuts, our appliances and electronics go dead, and a few minutes or a mere few hours later, everything comes back on. The most inconvenience we suffer is having to go around the house and reset all the clocks.

The longer blackouts last, the greater the chance of disaster. Food spoils. It becomes harder to prepare meals. People can't get to their jobs or provide well for their needs. If they don't have a landline, they can quickly lose touch with family and friends. ATM machines don't work, so cash is tight. (This is another reason why you should always keep a little cash on hand.) Gasoline is scarce because gas station pumps lose pressure and stop working altogether. Depending on the season, a long blackout can quickly become life-threatening if people cannot heat or cool their homes and businesses. They can't check the news and lose contact with the outside world. In some cities around the world, blackouts become the flashpoint for civil unrest as looting and rioting ensue.

All because the lights went out.

Getting through a blackout—big or small, domestic or foreign—is all about knowing what to expect, and how to prepare.

Some thoughts:

Surviving a Blackout in Your Home

- If you lose power while at home, you have several advantages on your side. You are not surrounded by strangers, all of whom have differing notions of what to do. You are the lord and master of your castle—as long as your kids and significant other agree! You know the space. You can find a lot of helpful tools and equipment in the dark. At the very least, you know where the flashlights are. With these in hand, you can tend to the needs of your home and family.

- As each season passes, take some time to lay in supplies that can help you if you had to live through a blackout of some duration in your own home. I would absolutely tuck away flashlights by your bedside and in several rooms of the house, and stash fresh batteries nearby. Be careful when using candles because they can be a fire hazard. Flashlights are better. Headlamps—the kind that fit around your head—are great because they keep hands free and allow you to assist others.

- If you haven't already, invest in an emergency hand-cranked USB charger. It takes a good bit of effort to hand-charge your mobile phone, but then you can check local news, weather, and municipal bulletins from your phone. Alternatively, a hand-cranked emergency radio (or one with solar power) can also be handy, especially if the blackout is far-reaching and local news outlets are impacted.

- Learn the ins and outs of your home so nothing surprises you if lose power. Electric stoves will obviously not work. Nor will ones with natural gas unless you can safely light the burners. You won't have hot water if you have an electric hot water heater. If your water is supplied by a private well, you might very well lose water pressure or access to water entirely. Be prepared to fill a bathtub with water *before* predicted storms.

- Every region of every country has a different blackout risk and response time. If you live in a part of the world where blackouts are likely to be days long, consider investing in a backup natural-gas generator for your home. (See my advice and caveats on these in the section on Blizzards, page 116.) A small solar generator might be fine, if all you need to do is power small electronics like mobile phones, computers, or a toaster and microwave. Small solar generators can also power small camp refrigerators. You can find more powerful solar generators that will run your refrigerator for hours at a time, but they are expensive and when the battery runs down, you will need to unplug them from the appliance and charge them from an outdoor portable solar panel. Decide what makes sense for you and family.

- If you must use a gasoline-powered generator, be aware that in storm conditions, even if you place the generator outdoors, wind could blow carbon monoxide back into open windows or doors. You must keep these and all other openings to your house shut when those types of generators are running.

- Minimize opening your fridge or freezer. The fridge will keep food safe for 4 hours without power, the freezer for 48 hours if it's jam-packed.

But opening the doors quickly decreases the appliances' ability to keep food safe. If you have advance warning, you can fill containers with water, and store them in the freezer to keep items cold. During a blackout, consider eating out of the pantry to minimize opening the fridge. Always assess the freshness of your food before consuming it after the power returns.

- Life-threatening blackouts often occur in the extreme seasons—winter and summer. I've covered most of my advice on *winter* blackouts in the section on Blizzards, page 116. Blackouts that occur in warm weather can be devastating because many structures—homes, businesses, apartment buildings—are not built to function well without air conditioning. In many buildings, you cannot open, or easily open, windows to admit a breeze or to encourage cross-ventilation. If this is the case, draw the blinds and curtains to shade the house from the sun. Do this as soon as possible, before the house heats up.

- Heat exhaustion, the beginning of *hyperthermia,* is marked by a red face, sweating, and extreme thirst. If unchecked, the victim's condition progresses to heat stroke, marked by a high body temperature, a weak heart rate, nausea, discomfort, and no sweating. This is extremely serious, and could easily cause death if left untreated. Children and older people tend to suffer the most. The quickest, smartest thing to do is move them to a more comfortable site.

In extreme blackout situations, many cities set up cooling centers where citizens can come for hours just to stay safe. If this is not an option, you can move to a shopping mall, library, or any other location in town where air conditioning is known to be operational. If you can't do that, consider moving all sufferers to a floor in the dwelling or apartment that is not as hot. (Heat rises, so upper floors of houses and apartment buildings are hotter.) Get the person to drink plain water to stay hydrated, and help them to cool down with cold, wet towels around the neck, wrists, armpits, and ankles. A cold bath or shower may help, but you may have to assist the person so they don't hurt themselves. You could wave a towel to manually "fan" the person. If the temperature in your home is higher than 95 degrees Fahrenheit (35 degrees Celsius), avoid using an electric fan if the power comes back on. At high temperatures, fans are only blowing hot air around and can actually cause the victim to overheat.

Surviving a Blackout
in the Big City

Urban areas present significant challenges during blackouts. In winter, tall apartment buildings might retain warmth a little longer than single dwellings. But in summer, a landscape that favors asphalt, glass, brick, and steel quickly become uninhabitable without some way to cool living spaces. Everything feels like a bigger, insurmountable problem. The water's out because tall apartment buildings rely on electric pumps to send water to the roof, then down to the various apartments. When the pumps go, so does the water. Transportation systems are disrupted, so people can only get around on foot, in private vehicles, or via buses—if they're running. Elevators and escalators stop working, making it hard to get around within a building. Walking down several flights of stairs is physically demanding for many people, and if they manage to get to street level, getting back up is virtually impossible. I had a friend who once had to evacuate an office building with her work colleagues. They trudged *down* what may have been 40 flights of stairs, and the next day everyone in the office complained that they couldn't move!

- If the city is your home, as soon as you get settled in your neighborhood, investigate all the potential services that have historically been made available to citizens during past blackouts and extreme weather events. Libraries, shopping centers, and senior centers are typically places used as shelters because they may be equipped with backup generators. You'll want to keep a printed list of municipal phone numbers handy, or pre-loaded onto your phone, so you can access it immediately during a blackout and call for information. You don't want to be looking for critical info in the dark, and you certainly won't be able to surf the web.

- City grids are subject to enormous demands for power. And they can be *overtaxed* at certain times of the day. It's wise to prepare for surges ahead of time with surge suppressors on your expensive electronic equipment. During an outage, you might want to unplug those pieces until you know the power is back for good.

- Well before a blackout, you should know your evacuation route in your home building, and in any buildings where you are staying temporarily. Not all staircases in large city buildings descend to the street. Know which is which before you open that fire door and begin walking down. Mistakes like this are annoying, not to mention exhausting.

- Any time you're riding an escalator or a moving walkway (like the ones found in airports), it's wise to hold onto the handrail. I understand why people don't. But if you're not going to hang onto the thing, at least stay within reach so you can glom onto it if the power fails. Be aware that even if you do grab the rail, not everyone traveling behind you will have the reflexes of Spider-Man. Stay alert, and be prepared for the possibility that a sudden stop will send someone flying into your back.

- In a city, checking on your neighbor is often a matter as simple as walking down the hall and rapping on the door. Be that kind of neighbor. Check on the people you know need a little special attention, especially the elderly. Just knowing some of the basic tricks I've given you—how to stay warm or cool—just might help you save a life.

Surviving a Blackout on Trips Abroad

You never know when a business trip or vacation will be horribly disrupted by a blackout. In most developed parts of the world, blackouts don't last long, but they can be extremely widespread. In June 2019, the power went out in three South American countries—all of Uruguay, and parts of Argentina and

Paraguay. The power was restored the next day, but that single event impacted *48 million people*. That's enough people to fill five and a half New York Cities.

If you're leaving for a trip abroad, you really ought to be checking the weather before and during your trip. You might want to consider under what conditions you will cut bait and simply fly home. An awesome vacation in an exotic and remote locale may well mean that food, water, and emergency services will be severely impacted during extreme weather events and blackouts. And if you flew somewhere for a conference or business meeting, you have to wonder how productive that venture is going to be if no one can see the PowerPoint presentation you brought.

As I say, in most cases, the power will be restored, and you'll be back to whatever you came to do. But if you're in a locale where food, water, and essential services are impacted, you might be better off switching to Plan B, C, or D. It is always wise to travel with the phone numbers and contact information for your home nation's embassy. Protect yourself with refundable/exchangeable tickets, travel insurance, and medical evacuation and repatriation coverage. You shouldn't have to lose your money, or your life, if you're stranded somewhere because the lights went dead.

PANDEMICS

A nd so we arrive at Frankenstein's monster. The boogeyman of all boogeymen. The zombie apocalypse. The germ of germs. The bug of bugs.

The world's experience with the COVID-19 coronavirus may have seemed to come out of left field, but pandemics—diseases that attack an entire swath of a nation or the world—have unfortunately long been part of the human experience. We had them in the Middle Ages. We had them in the Renaissance. We had them at the time of the American Revolutionary War. We had a massive one just after World War I that got one-third of the world sick, and claimed the lives of 50 million people, 675,000 in the United States alone. Of all nations, India lost the highest number of people—14 to 17 *million*. That's just the number of people we *know* who died of the so-called "Spanish flu." Modern scientists and historians think the number was much, much higher.

So: When it comes to pandemics, the question is not so much why are we having one as, why *wouldn't* we have one?

A terrible truth of pandemics is that we only seem to pay attention when the numbers of the sick and dying are catastrophically large. I'd venture to say that everyone has heard of the 1918–20 pandemic because it's the one the media talked about constantly when COVID-19 leapt to the world stage. It was the most severe pandemic of the 20th century. It was an H1N1 virus. If that term sounds familiar, it's because the world experienced an H1N1 outbreak in 2009, which goes down in history as the *first* pandemic of the 21st century.

Bubonic plague—aka the "Black Death"—is legendary because a person can hardly study world history without knowing that Europeans in the 14th century died in droves—75 million to 200 million is the estimate—from a disease that "came" from rats. But the bacterium that causes that horrific "historical" disease is actually carried by fleas, and it *still* exists in parts of the world today, including places in the United States.

So, in a way, when it comes to pandemics, what's old is what's new. And unluckily for us, pandemics are likely to become more common. Why? Let's think about that for a second.

The bubonic plague spread the way it did because it was the first time in history that the world had opened up to robust mercantile trade by ships and land routes. The 1918 pandemic is believed to have originated in New York and spread worldwide because so many troops were traveling in tight quarters as a result of World War I.

In the modern world, humans are far more mobile than medieval merchants or World War I troops ever were. Before and after COVID-19 locked the world down, businesspeople and tourists traveled the globe at the drop of a hat for meetings, conferences, wonderful vacations and adventures. If they pick up a bug on their trip, they can easily transmit that pathogen to many more people when they land 24 hours later in their new city or country.

Many of history's most deadly diseases—influenzas, malaria, measles, plague, smallpox, tuberculosis—were originally passed from animals to humans. In 1918, the world population was about 1.8 billion. Today, ours is a teeming planet of more than 7.8 billion people. At no time in history have so many people lived on Earth. That means we have more mouths to feed than ever before. Feeding that appetite forces millions of people to interact daily with tons of domesticated animals, which are raised and slaughtered for our food. Even today, it is our unlucky lot to "share" numerous diseases with cattle, sheep, pigs, and chickens. (By one estimate, about 10,000 animal viruses could conceivably infect humans.) In some countries, horses are raised for food, and in others for companionship and equestrianship. Where all these livestock congregate, rats and bats are often fated to be. They are certainly two creatures known to pass communicable diseases to humans. To make matters worse, humans routinely harvest bat guano (i.e., excrement) to make fertilizer. Given this menagerie and our interactions with the animal world, it's not too unreasonable to ask: How are we not getting sick more often?

Numbers don't lie, nor do science and history. In 1918, the virus did eventually die out and life returned to normal. That was the good news. The bad news was, it came back—twice—and killed *far more* people than it did on its first outing.

In a natural disaster, people are accustomed to relying on help from outside the realm of the epicenter. If a tsunami strikes Australia or the Philippines, the people in those countries know that they can rely on help from nations throughout the Pacific. If a hurricane devastates your city, you know you can count on help from nearby cities and states, not to mention the federal government. Volunteer first responders stream in from every part of the nation or globe to help.

But pandemics throw that equation out the window. They impact everyone. If every city or country is on lockdown, who can send help? If every corporation or non-profit aid organization is strapped for able-bodied employees and funds, how can they send help? Closer to home, if people in your town get sick and essential services are interrupted, how will you get by?

It absolutely makes sense for all families to be prepared for long periods of self-sufficiency and isolation. If I had told you a decade ago that you needed to prepare your family and home for an emergency that could last 2 to 6 weeks, you would have said, "Yeah, sure, that's wise advice, Doc. I really should do that … *someday*." And you'd go off and postpone those actions for another day.

I'm not pointing any fingers. It's simply what we do as humans. We are hard-wired to devote more energy to present concerns than to hypothetical disasters that may possibly happen in the future. It's the same reason many people don't save very well for their retirement.

We're back to the old story of the ant and the grasshopper. Hands down, in the human psyche, the demands of the present always outweigh the demands of the future.

Individuals do this, and so do organizations. I witnessed this firsthand in the military. In nearly every simulation the analysts ran, they'd discover that we weren't prepared. We did not have *enough* supplies or the *correct* type of supplies. The food, medicine, and equipment we stockpiled in government warehouses for emergencies had either expired, or were in serious need of maintenance. The experts would bounce this sorry state of affairs up the chain of command. The chiefs would act for a little while, until budgets got tight and the needs of the present again outweighed the disasters of the future. And then the chiefs gave it a pass.

But all of us know better now, don't we? We don't just know things can get bad; we've *lived* through what one version of the bad looks like. Millions of people lost their jobs, and thousands of firms went belly-up.

At the same time, during the COVID-19 pandemic, lots of people, businesses, and organizations learned how to "virus-proof" themselves. They got leaner, smarter, and more adaptable. I would argue that municipalities, medical firms, and hospitals are better today at knowing what they need to do to serve their populations. Of course, getting the supplies they need is a different story.

And you? I am sure that you learned a few things about how to protect your family. By virtue of what you have experienced, you are already halfway there.

With this in mind, I propose to arm you and your family with thoughtful guidance on what you can realistically do or assemble to prepare for another zombie apocalypse.

But before you read on, I'd love it if you could sit down with a sheet of paper and come up with a list of your own. Think back to the spring of 2020, when COVID-19 quarantines began around the world, and ask yourself: If you knew then what you know now, what would you do differently? What would you do to prepare your family? Your home? Your job or business?

When my wife and I did this recently, we came up with the following list. Admittedly, it is ambitious, and touches on many of the points I've raised earlier in this book. But I think it's the best way a family can be prepared to hunker down, stay safe, and ride out future pandemics.

Medical Preparedness

1. From this moment forward, resolve to stop deferring medical treatment you know you or a family member needs. Doctors' and dentists' offices were slammed after the quarantines lifted because so many people had been forced to put off critical care, elective surgeries, or other serious health issues during lockdown. There's a reason that every major airport in the world has a dental care facility near the terminal— experiencing dental pain for any length of time is excruciating. Don't delay this or any care. Make a pact with your spouse or loved one to nip any family medical issues in the bud when they are first noticed. You never know when your access to medical care will be curtailed by a quarantine. If you must visit a doctor's office during a pandemic, call them first and ask them to explain their protocol for office visits. They may want you to announce your arrival via text, and to remain in your vehicle until they summon you.

2. Make a plan to keep all family prescriptions current, and see that you maintain a 30-day supply of all your medications in your home. See if your pharmacy has an online mail-order delivery option. If so, activate that online account in case you are not able to visit a pharmacy in person to pick up what you need. Get your supply of contact lenses, reading glasses, or the spiffy new prescription glasses you want. Stock up on hearing aid batteries. Join the patient portal operated by your

primary care physician, so you have a secure way to communicate with your provider without leaving the house. Many health care providers jumped headfirst into the realm of "telehealth"—online video consultations—during the COVID crisis. Ask your doctor how you can sign up for this service. Not every medical situation requires a physical visit to a doctor. That way, in case of a lockdown, many of your family's minor health issues can be treated by a short conversation online with your doctor.

3. Upgrade your home's first-aid kit. It's fine to have a stash of Band-Aids in assorted sizes. But it's wise to look further afield. Consider assembling the kit I presented on page 20.

4. During a lockdown, if you or a family member believe they may have contracted the illness du jour, your first instinct will be to contact your doctor. Doctors and nurse practitioners will be wondering if your family member is running a fever, and what their oxygen level is. You can answer that question if you have both a digital thermometer and an oximeter, which is that small plastic device doctors clip to your index finger during any routine exam. You can purchase an oximeter from most online retailers for $30–$60. A normal reading is 92 or above. The higher the altitude where you are taking the reading, the lower the reading will be. Any reading under 90 is concerning.

Special Considerations for Babies and Elders

1. Make sure you have all the supplies you need to care for these two specific demographics in your households. Babies require diapers, wipes, specific foods, prescriptions—as do some senior members of the family.

2. Assess all medical devices or equipment to be sure they are in good working order. Devices such as CPAPs can run off some types of battery-powered generators in case of a power outage; investigate what makes sense for the equipment in your home.

Medico-Legal Issues

No one likes to contemplate their mortality or that of the ones they love, but now is the appropriate time to execute or update your will, advance directives, health care powers of attorney, and so on. Make sure that you have authorized a trusted individual to access your medical information in an emergency, if necessary,

Healthy Living Choices

1. Home exercise equipment experienced a renaissance during the COVID quarantine, and for good reason. Buy, service, or replace all the equipment you think you might need to stay healthy while sticking close to home. Not everyone needs an exercise bike or free weights. Often a set of moderate hand weights or resistance bands, partnered with socially distant walking or an exercise app on your phone, can keep you in excellent shape.

2. People with health problems and heavy smokers are easy targets for any airborne virus. Now is an excellent time to quit smoking or embark on an exercise program that will support your cardiovascular health.

Tools, Supplies, Kits

In addition to a first-aid kit, every home should have a couple of classic "survival" tools that will make life easier in the event you lose power, or need to fix something in an emergency. I've spelled out the critical components of a robust bug-out bag on pages 55–60. But during a pandemic, you won't be bugging out as much staying in. In this case, your critical gear should include:

1. Crank radio: The classic hand-powered radio will allow you to tune in to AM/FM stations, and hear critical weather broadcasts from the National Oceanic and Atmospheric Administration (NOAA). Yes, you can find ones that include built-in flashlights and ports for charging smartphones, but bear in mind that such power-hungry features will force you to crank endlessly just to send a text or find

your way in the dark. A solar-charged device might be more suitable to your needs.

2. Stock up on flashlights and a good supply of batteries. While you're at it, assess the battery and power needs of all the critical devices in your home. And yes, if you have a large brood, the TV remote definitely counts as critical equipment. If you have an old VCR and videotapes, you might want to rethink tossing them out.

3. A good "survival" knife: You don't have to go Crocodile Dundee on this item, but you'll want a decent knife for important tasks such as slicing duct tape or cutting rope. Some small folding knives perform both jobs handily. A "combo" blade will have a partially serrated edge.

4. Tool kit for minor repairs. In a pandemic, you are unlikely to want handymen or other service people over to your home to do minor repairs. If you don't already have a small tool box, you might be able to get by with a multitool. They unfold into a dozen different tools, and are small enough to tuck into a pocket. Be sure your first-aid kit has its own pair of scissors, so you don't have to go searching for a pair of household or utility-drawer scissors to treat someone in an emergency.

5. I love duct tape. Trust me. It's infinitely adaptable, and strong. Get a roll and tuck it somewhere handy. The MacGyver in you will feel immeasurable delight if you ever have to use it.

Stocking Your Kitchen and Pantry

1. Food breaks down into two basic categories—perishables and non-perishables. Yes, you can certainly grocery-shop a few days before a lockdown goes into effect, but most of the food you store in your fridge and freezer won't last long when you're cooking three meals a day, seven days a week, for the whole family. You can choose to suit up in protective gear to grocery-shop on a regular basis *during* lockdown, but if you're like most people, you will want to minimize trips out of the home. The wisest course is to stock up and learn to live out of your pantry. (See my suggested shopping list on page 233.) Canned or

dried beans, tinned fish and meats, and peanut and other nut butters are always smart sources of protein. After that, you will want to store wheat or gluten-free pastas, rice, oats, tomato and other sauces, canned vegetables, condiments, and so on. Buy ingredients you *know* you will enjoy and that you *already know* how to prepare. It makes no sense to buy three types of flour and yeast if you've never baked a loaf of bread in your life. Be sure you also have an adequate supply of cooking oils. (Those stored in metal containers last longer than those in glass, and glass containers are better than plastic.)

2. Go through your pantry right now and discard any food items that are dangerously past their expiration dates. Assess your storage needs and figure out how to optimize the space.

3. Inspect all your cooking utensils and tools to see if any need to be replaced. Replace iffy can openers. You don't want to be without one when you're staring at a pantry full of cans and no desire to suit up for a trip to the grocery store. If you need to leave your home in an emergency, you can take along a portable, military-style P51 (smaller) or P38 (larger) can opener, which attaches to a key chain. Consider purchasing one of these for your car's glove box or bug-out bag.

Your Quality-of-Life Needs

1. Every family needs recreation. Think board games, puzzles, books.

2. Every parent needs their sanity restored in the face of the 24/7 presence of their kids. Consider noise-cancelling headphones or earplugs for when you absolutely need to tune out.

3. Consider having 2 weeks of cash on hand, some in small bills and stored in a safe place, in case you need to tip delivery people, pay for cash-only services, and so on.

Your Personal Protective Equipment ("PPE") Disinfecting Station

1. Apartment dwellers typically have only one way into their home, but in detached dwellings with multiple entryways, it's smart to designate one door to serve as *the* entrance your family will use during a lockdown. As soon as anyone enters, they can shuck off coats and shoes, "wash" their hands with sanitizer, and deposit any packages, keys, or belongings on the floor or a low table to be disinfected prior to bringing these items into the house.

    You may also want to keep a boxcutter tool for opening and breaking down shipping boxes in this area (on a higher surface if small children live in your home), and a bag or box for filling with these collapsed containers to be promptly recycled. Store all critical personal protective gear—masks, gloves, sanitizer, spray disinfectants, alcohol wipes—in this location. Supervise small children so they don't linger in this area and re-contaminate themselves, their siblings, pets, or you.

2. The nation's store of cleaning supplies was severely limited during the COVID crisis. It's smart to maintain a decent stockpile of anything you need to clean your home on a regular basis, such as paper towels, dish soap, dishwasher soap, hand soap, disinfectants, mops, disposable sweepers, etc. For 2 to 6 weeks, a family of four could theoretically use about four to five bottles of hand sanitizer. Use yours sparingly. After the first few days of lockdown, when your family is getting used to the routine, you'll only need to use sanitizers if you run errands out of the home. Be thoughtful about where you store your supplies. Many cleaning liquids and compounds are sensitive to heat, including the temperature in a summertime garage, laundry room, hot water heater closet, or similar. Sensitivity might mean inflammability, evaporation, or degraded efficacy.

3. Stock up on essential toiletries such as laundry detergent, deodorant, soap, shampoo, personal products, razors, feminine hygiene products, and, yes, toilet paper. Just as a reminder: "Cleaners" get rid of dirt and grime; disinfectants kill germs. I'm a fan of bar soaps because

they will dry "clean" and can be used in a variety of ways. As we've seen, millions of other people will be thinking the same thing, so buy responsibly. A family of four will use about 17 rolls of toilet paper in 2 weeks, not 300. (Unless you have other issues!)

4. It's smart to lay in a good supply of face masks, even if you plan *not* to leave your home for the duration. You simply never know when or if an emergency will force you outdoors. Many people make their own masks, which is certainly fine, but the best protective face gear are N95 "respirators," so called because they filter out 95 percent of airborne particles and can be molded to fit closely to the face. I do not discourage their purchase, but since they will be in high demand for first responders and medical personnel, please consider donating yours if you learn of a desperate need for them in your community. N95s should be discarded after any high-risk events, if they no longer fit well, or if they become obviously damaged, wet, or dirty. Homemade fabric masks, which filter about 60 percent of airborne particles, should be laundered after use. Efficacious masks have two to three layers of protection.

5. If someone in the household is likely to be designated an essential worker, take the time now to discuss how you will support this person's comings and goings while maintaining the safety of everyone else in the home. Some spouses I know choose to sleep in different quarters, or set up zones in the home that they alone use. Not every family is going to be able to do this, but it's worth taking time before an emergency to discuss your options.

6. Have a plan in place if someone in the household needs to be isolated. Ideally, they would have their own sleeping space and bathroom. One other person in the household needs to be designated as their caretaker, and be supplied with the appropriate PPE for the job.

Traveling During a Pandemic

Like many animal species on the planet, humans are social creatures. We love hanging out with other people. We hug them, kiss them, share food and drinks, and chat them up within inches of their faces. That's how we feel close to each other.

It's also how we spread germs.

If that weren't enough, when we're not hanging out in the presence of others, we're touching our own faces constantly. One study found that people touch their faces twenty-three times an hour. This behavior is practically involuntary. We do this the way we blink—without thinking.

Doctors are trained from their earliest days in med school to hold their hands at their sides after touching patients. Job One is washing their hands before seeing the next patient, meeting with staff, or performing a single administrative task. We also learn not to touch our face in course of a normal work day.

If you must travel, or choose to travel, during a pandemic, you have to harness your inner germophobe. Before leaving your home, pack masks, hand sanitizer, and sanitizing wipes in your carrying bag so you have easy access to them on your trip.

The big decision you will most likely have to make is whether you will fly or drive to your destination. Flying takes less time, but on every step of your journey you will be in close proximity to other people. Driving takes longer, but you'll be in the relatively safe environment of your personal vehicle. The catch is, you'll still have to stop often—to pump gas, use rest rooms, buy food, and possibly check into hotels along the way. Travel by train or bus is a mix of both. They're not as swift as air travel, but you'll be spending a lot of your time in public spaces.

Obviously, the safest thing is to stay home, but not all of us have that luxury. Life doesn't stop during pandemics. If anything, it accelerates, complicates, exacerbates! There are older relatives who need help. There are kids at school who need to be retrieved and brought home. And there are work demands that require travel.

Plan what you're going to do before you do it. Whether flying or driving, the fewer tasks or stops you make, the better. Wear masks and gloves, and wipe down every surface before you touch it.

If you're driving, bring along as much food as you can to eat in your vehicle. Curbside pickup at restaurants along your route is safest, takeout is second-safest, and sit-down dining the least safe. If you absolutely must dine indoors at a restaurant, choose seats or booths away from others and wipe down your seats, tables, and even utensils before using. Don't worry about looking like a freak. Anyone who's traveling during this time should be doing the same thing, and the places you are patronizing ought to be grateful for your business. Place your order all at once to reduce frequent foot traffic and interaction with your server. Ask that they not bus the table until you have departed. And whatever you do, don't forget to wear a mask each time you leave your vehicle.

If you're flying, book window seats when you are buying your tickets. That will reduce your interaction with others. Bring snacks in your carry-on bag. Practice good social distancing as much as possible. In airports, and in bus or train stations, you want to scope out a seat away from other travelers. Limit bathroom visits, and the whole time you're in that bathroom pretend you're a surgeon—don't touch yourself until you've had a chance to scrub down.

Yes, traveling like this is an absolute pain. I traveled cross-country by plane, from New York City to Colorado, during the COVID crisis, and it was definitely not fun. I blew through a lot of gloves, wipes, and disposable masks. But I saw no other option. I was anxious to get home to my family. If I had not had access to a fair amount of PPE, I would have had two options: postpone the trip until I could stock up, or wash the gloves and masks along the way. In the end, the choice to go or not go shakes down to a judgment call. Psych yourself up to follow through on your decision.

It might be wise to apply for the Trusted Traveler program that is most relevant to your typical flight destinations. This will allow you to reduce the amount of time you must spend at airport security checkpoints. Before you depart, make sure you investigate whatever travel quarantine restrictions the state or nation you are flying to has put in place, as well as the state or country you are coming back to.

PART III

HUMAN-CAUSED DISASTERS

GAS LEAKS

’ve had more gas leak scares than any human alive, but thankfully not a single one of them ended badly. I’ve had contractors operating a backhoe in my front yard hit a natural gas line that was poorly marked and then run for the hills, terrified for their lives. I’ve had my daughter wake us at 2 a.m. on a winter’s night because the overpowering odor of natural gas was filling the house. Our entire family hunkered down in our car in the driveway until the fire department could come check it out. Another time, a carbon monoxide detector in a home we had rented sent us scurrying out to the car again, this time at 3 a.m. Turns out, the detector malfunctioned. But at least we were safe.

You don’t want to mess with gas leaks. The symptoms of natural gas exposure seem mild at first. In short, a victim may behave like someone who has drunk too much alcohol: a little tipsy, unsteady on their feet. Then comes nausea and throat irritation. Finally, they have trouble breathing, pass out, and die.

Natural gas, which fuels many of our appliances and keeps our water and homes warm, is surprisingly odorless. The rotten egg smell that we associate with natural gas is an artificial additive; it is introduced into the gas by local gas companies so that humans can detect leaks by smell alone.

You get no such warning with carbon monoxide (CO). It’s an odorless, tasteless, invisible by-product produced by the incomplete combustion of petroleum products. It can leak from any fuel-burning appliance (or vehicle) in your home. For some unexplained reason, the hemoglobin molecule in human blood is attracted to CO and binds to it, slowly reducing our blood’s ability to bind with and transport oxygen. There’s no evolutionary advantage to this mechanism, but the results are deadly.

Recently I’ve done a number of news reports on families who have been poisoned by carbon monoxide. In the emergency room, I have interviewed entire families who complained that they all got headaches every single night before bedtime, only to have those headaches disappear the next morning when they went to work or school. The culprit is usually discovered to be

CO. The families who live to complain about such symptoms are the lucky ones. The more common scenario? A family goes to sleep, a leak occurs in the night, and they don't wake up.

In the last decade or so, homestay rentals have been hailed as the hot affordable lodging option around the world. But the mom and pops who operate these places may not always install carbon monoxide detectors. These devices are the only way you can know if your living space has been infiltrated by CO.

Natural Gas Tips

- Any time you move into a new home or apartment, take note of which appliances burn natural gas and be prepared to sniff out leaks, at least for the first couple days in the home. The previous owner may have become so accustomed to that sulfurous smell that they didn't flag it as a concern.

- Day or night, if you do smell natural gas, resist the impulse to flick on the lights as you go about your investigation. Battery-powered flashlights are safer. Even a little spark in the circuitry behind a switchplate can cause an explosion.

- Get out of the house first, *then* use a mobile phone to summon help. Some natural gas companies express concern that mobile phone use may ignite gas.

- Natural gas is *lighter* than air and rises. As you leave the dwelling, crouch close to the ground, the way you would to avoid smoke in the event of a fire.

- Be aware that some of your family members may already be suffering from exposure. Be on the lookout for groggy, sluggish behavior, and be ready to assist them to evacuate. They should perk up once they get clear from the house and breathe fresh air. If they do not, seek care for them immediately.

- Allow no one from your family (or workplace) to re-enter the site of a natural gas leak until the area has been inspected and cleared by the local fire department or representatives of the gas company.

Carbon Monoxide

- Install carbon monoxide detectors as soon as you take possession of a new home or apartment. Carbon monoxide neither sinks nor rises. It's only slightly lighter than air, and tends to mix with the surrounding oxygen. Ideally, you'd install CO detectors on every living level, in your basement, and outside each bedroom. CO-only detectors are installed at knee height. (Smoke detectors are placed on or near the ceiling.) If you're using a "combo" alarm, install according to the manufacturer's instructions.

- If you have a garage attached to the house, you want to install a CO alarm about 10 or 15 feet (3 to 4.5 meters) away from the garage because car exhaust is heavy with carbon monoxide. The best place is usually in the interior hallway leading to the garage entrance. Never install the detector outside the garage; it would be utterly useless in the open air.

- Cars with keyless ignition may require caution when exiting the vehicle. There have been cases where people have been distracted when they arrived home. In their haste to go about their next errand, they unknowingly left their vehicle running in their garages, erroneously believing that they had shut off the ignition. Carbon monoxide built up in the closed environment of the garage, penetrated their living spaces, and killed them. An "old-fashioned" ignition key demands attention; it has to be switched off or you probably can't use the other keys on your ring to enter your house. Keyless fobs are convenient because you can leave them in your pocket or purse. Auto manufacturers have only recently begun installing auto-shutoff in their keyless cars. If your keyless vehicle is an older model, you really have to train yourself to check that the car engine has stopped running.

- Check each detector's batteries on a regular basis, and record each unit's expiration date in your digital calendar so you know when to replace the detectors. (Yes, the devices themselves have a lifespan, usually indicated on a label affixed to the device.) And of course, change all the batteries on these units at the customary, recommended intervals—when you change the clocks each year for the start and end of Daylight Savings Time.

- If all your appliances are electric, and you have a detached garage, you don't need carbon monoxide detectors. But local codes may insist on their installation.

- If you're traveling and staying in atypical lodging such as homestays, B&Bs, etc., or traveling to a foreign country where detectors may not be mandated, it doesn't hurt to pack a portable CO detector in your checked luggage. Remove the batteries while flying, and reinstall them when you reach your destination.

When Hospitals Are Necessary

In most cases, the "cure" for the two most common gas leaks is fresh air. But if you've fled your home and someone in your party is still suffering symptoms, they need to go to a hospital. A blood test will tell if the gas molecules, for whatever reason, are continuing to flood the person's hemoglobin. Doctors treat this problem in one of two ways—either by administering oxygen-rich air, or, if the situation is dire, by placing the person in a hyperbaric chamber. In such chambers, the air pressure is manipulated to be three times higher than what is found in ordinary air. In such an environment, our lungs can inhale even more oxygen, thus overwhelming the tainted hemoglobin, and driving out the gas molecules. If the person doesn't recover, they can be left with stroke-like symptoms permanently.

You don't want to mess around with gas leaks.

Specialized Gases

A gas leak situation becomes problematic if the type of gas is unknown. All gases behave differently. In a freight train derailment in South Carolina in 2005, 120,000 pounds (54,431 kilograms) of chlorine gas was released into the atmosphere, killing nine people and forcing 5,000 people to evacuate their homes. It still stands as the worst chlorine gas disaster on U.S. soil, and accidents involving this "important" industrial gas have become unfortunately common. Chlorine is *heavier* than air, so in an enclosed space it drops to the bottom of a room. So you would never flee an indoor chlorine leak the way you flee a leak with natural gas, which rises. A home gas leak is unlikely to be something other than the two gases I've described above. But you may come in contact with rare or unusual gases at your place of work. In many work environments employers are required to post "safety data sheets" on potentially dangerous substances in the workplace, so they can be easily inspected and read by employees. If on-the-job training is not extensive enough for your taste, make it your job to familiarize yourself with the characteristics of each new substance *before* there's a problem.

HOME OR BUILDING FIRES

The first thing anyone does upon buying a new home or renting an apartment is take out an insurance policy to cover the dwelling and all their possessions. Your home is a place of shelter, a refuge from the world, a cocoon of love, and a place where family memories and traditions are forged.

But you don't feel any of that when you take out an insurance policy. It's all about how much they will pay you if the place goes up in smoke. So much to replace the structure. So much for the contents. And in return, how much you will be paying them annually to lock in this protection. So many of us sign on the dotted line without really contemplating what that type of disaster would be like.

But I have a different perspective on fires. It is gleaned from years of walking through hospital burn wards. The pain some burn victims experience is so indescribably horrific that there's no way any doctor can ever prescribe enough medication to quell it.

And the surprising thing is, the victims who feel pain are the *lucky* ones. They're the ones whose nerve endings are still functioning. They haven't been completely eradicated by the flames.

Whether in your home or anywhere else, you never want to be in a fire. There are simply too many variables, too many ways the situation can go from bad to worse.

For one, the smoke from fires quickly destroys visibility. People are not prepared to flee in such darkness, and it's often hard to gauge the complexity of the evacuation. If your home is on fire, you're unlikely to have emergency lighting, but you probably know exactly how to get to the front door and out of the house. In a theater, hotel, office, or business, your path might well be illuminated dimly by emergency lighting. But your escape will most likely be hampered by being in an unfamiliar environment or among a panicked crowd.

One of the most horrifying fires in recent memory was one that started in a Brazilian nightclub in 2013. It's believed the band that was playing set off fireworks or some kind of pyrotechnic display that set fire to the acoustic foam in the ceiling. Nightclubs are problematic anyway because they're usually kept loud and dark, but this one was worse. No fire sprinklers. No detectors. No exit signs. And only one way out. A recipe for disaster all the way around. It was the second-deadliest fire in the history of the country. In a flash, 245 people died, mostly because they couldn't find their way out and mistakenly congregated in the bathrooms. More than 600 were injured. That tragedy is a reminder that you should never enter a place of business and let your guard down without knowing how you're going to get out in an emergency. It's as simple as looking around and noting the potential exits.

Fires in buildings can often mean explosions. Blasts kill quickest because they're shooting shrapnel. Smoke inhalation is the next biggest—and quickest—killer. You really only have a few minutes or seconds to get out before the smoke obliterates vision and snuffs out your breath. That's what killed most of the people so tragically in the nightclub. What kills later are thermal injuries, skin burns, and damage to our very delicate human noses, eyes, and throats.

In both scenarios, homes and businesses, fires incinerate plastics, fabrics, and other substances that can release toxins into the air. These fumes are as dangerous as the smoke from the fire. There is never a good reason to linger longer than you must when a place is going up in flames around you.

To give you and your family the best possible chance in a fire, I urge you to do at least these four things immediately. You may not be able to control how your employers prepare for fires on the job. And you certainly cannot control the fire prep performed by the businesses you patronize. But you can absolutely lock down some best practices in the castle you share with the people you love most.

Please do these few things this week.

- **Check to see that you have smoke alarms on every level of your home. You probably do. And you probably have even gotten into the habit of changing out the batteries twice a year, the way we are always told to do. Great. (Changing the batteries when they beep, one by one, is really annoying.)**

But here's an extra step I'd like you to take: Check the expiration date stamped on each unit. Believe it or not, modern ones have them, because the sensors in each unit deteriorate over time and the entire device must eventually be discarded. (Most alarms contain a small amount of americium-241, which is a radioactive substance, so please dispose of them properly at a hazardous waste site.) On your phone's or computer's digital calendar, note the "replace by" date of each unit so you don't forget them. And when you go to replace the alarms, choose devices that all have the same expiration date, so you don't make yourself crazy.

If your smoke alarm does not have a "replace by" date, it should have a date stamped on its back that tells you when it was manufactured. Alarms should be replaced when they are 7 to 10 years old. If the plastic housing on your smoke alarm is a hideous yellow, I'll bet it's long past its expiration date. Go and check, and tell me if I'm wrong.

- See that you have enough fire extinguishers throughout the house, and that they are the correct type for home use. Experts usually recommend a multi-rated extinguisher that is capable of handling the three most common types of fires: Classes A (ordinary combustibles), B (flammable liquids), and C (electrical) fires. Some manufacturers label kitchen and garage extinguishers differently from other types. Choose accordingly. Store appropriately. These devices lose pressure over time, and need to be refilled or discarded. Because of this, it might make sense to work with a dedicated fire extinguisher supply company in your area that allows you to swap out devices when the time comes, rather than buy disposables at a big box store. Whatever your choice, note the expiration dates of each extinguisher on your digital calendar, so you won't forget them. Read and understand how to operate the extinguishers so you don't have to learn when you need it most.

- Teach your family how to properly operate a fire extinguisher. If you get your extinguishers from a local company, they may be able to provide you with a "practice" canister. Deploying these devices is best described by the acronym P.A.S.S., which reminds us to:

PULL the safety pin.

AIM the nozzle at the flames.

SQUEEZE together the two levers.

SWEEP the fire from left to right until extinguished.

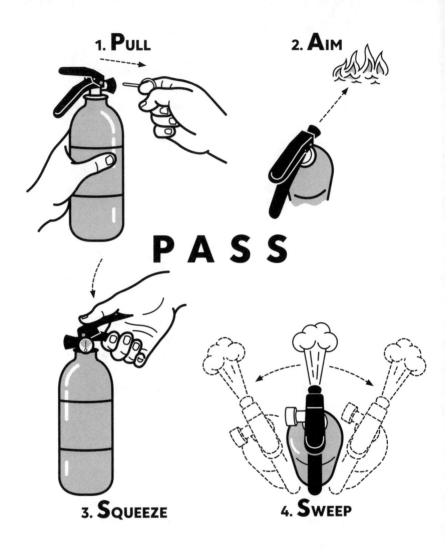

1. **Pull**

2. **Aim**

PASS

3. **Squeeze**

4. **Sweep**

We all know that we are supposed to make a fire safety plan with our families, but many of us never get around to it. Or if we do, it's a verbal plan—"Let's all meet outside by the big tree"—and we forget to do a drill so that everyone is on the same page. Please do so for real. It can actually be fun. I'd do a drill one night after dinner and before the family starts hanging out in front of the TV. No one will think it's fun to do a drill at 3 a.m.—especially your significant other!

Please note that if there is only one way down from the sleeping areas of your home, it might make sense to outfit one or more bedrooms with a two- or three-story escape ladder. These are small and can often be tucked under a bed or in a closet.

Surviving a Fire in Your Home

Practice good fire "hygiene." Never leave candles and flames unattended. Even a backyard grill can become a raging disaster if you're not careful. Stoking a backyard fire pit feels luxurious on a crisp fall night, but you really ought to drench the thing with water at the end of the night to be absolutely certain that it's dead cold before you and your guests turn in. If you're going for a special ambience for dinner or parties, I recommend battery-operated candles. The quality of these have gotten quite good, and the wicks mimic the warm flicker of real flames.

As a general rule, your best shot at putting out a fire is when it flares up in front of you. In a kitchen, garage, laundry, or small garden shed, you should only have to take a few steps to grab a fire extinguisher and quell a nascent flame. Never toss water on a kitchen grease fire; that will only make it worse. If a smoke alarm goes off in the middle of the night, chances are the situation is already out of hand and you ought to evacuate immediately.

Smoke rises and quickly obscures your way out. If you smell or see smoke, cover your face with a wet towel or washcloth, crouch down closer to the floor, and breathe through the wet cloth as you make your way out of the house.

Oxygen feeds fire. If you encounter a closed door on the way out, touch the surface of the door with your palms. If the door feels hot, don't open it. The fresh oxygen will exacerbate the fire. Seek another way out, even if you must leave by a window.

- If you are forced to exit from a window on an upper floor and you don't have an escape ladder, never jump to the ground. Two better options: You can hang onto the windowsill and dangle yourself out so you are closer to the ground, then release and drop the rest of the way to the ground. Or you can toss pillows, blankets, clothing, or even a mattress (if it is small enough) out the window first, and use them to break your fall.

- Use your escape time wisely. Parents should assist children, seniors, and anyone else in the home who needs special assistance.

- The time of day when a fire occurs will impact your response. At night, you can safely assume all family members are asleep in their beds. But a fire that occurs on the weekend, in daytime, may complicate matters because family members may be scattered throughout the house. Hammer out a divide-and-conquer strategy with your spouse now, *before* there's a fire emergency. For years, I had the following protocol in place with my wife: At the first sign of a fire, I would run through the house and physically corral the kids. My wife's job was to immediately leave the house, text the kids—because every kid on the planet is glued to their phone—to evacuate, and *then* phone 911.

- Besides developing an escape plan with your family, it helps to ask your children from time to time, "Hey, if you were in this room and you heard me yell 'Fire!' how would you get out of the house?" Do not be afraid to ask that question dozens of times in different rooms of the house. I have found that even little kids appreciate being asked to participate in making a serious, obviously grown-up plan.

- * When you are finally out of the dwelling, you need to be as cold and impersonal about your possessions as that insurance policy of yours we talked about earlier. Nothing you own is worth the loss of human life. Under no circumstances should you ever go back inside a burning house to retrieve valuables, documents, computers, hard drives, or anything else. Focus on getting your family—and pets—out of the house safely.

Surviving a Fire in an Office or Commercial Building

Offices, stores, malls, hotels, theaters, and other commercial structures in most developed nations will have illuminated exit signs, backup lighting, fire stairs, fire sprinklers, and evacuation maps. Any time you enter a new structure, or check into a hotel, make it your business to check for these signs and exits. In the back of your mind you should be asking yourself the same question I suggested asking your kids: "If I had to get out of here right now because of a fire, how would I do it?"

- Major hotels should have an evacuation map posted on the interior of your hotel room door. But I have checked into many hotels, homestays, and B&Bs worldwide where such maps were nonexistent. I love the thought of a charming getaway as much as the next person, but please take the time to figure out your own escape route if it is not provided for you.

- Elevators are unreliable in times of fire. Don't enter them. Look for the fire stairs instead. In certain commercial buildings, only certain stairs are marked as fire exits; they may well be the only stairs in the structure that exit directly to the street. Other staircases may only be designated for the use of firefighters. Follow the posted instructions. Trust me. The next time you're in a strange building, look around. Stairs are not always in the first place you'd look.

- Your place of work should hold regular fire drills, with one office mate designated as a fire marshal, charged with flushing everyone out of the building. Participate in these drills with your full attention. They may well seem pro forma, but you never know when you'll need to call upon what has been drilled into you. Who knows? You might even get to be the one who wears the cool yellow vest!

- Maintain situational awareness. Keep your eyes off your phone and your head in the game. You never know if the situation you *think* is a fire is actually something far worse. In certain parts of the world, the fire alarms may have sounded because you're in the midst of an earthquake. There may be a gas leak.

- Stay alert even when you get onto the street. I'll talk about this in another section, but terrorists have been known to phone in bomb threats with the express purpose of flushing people out of office buildings, so they make an easy target for a secondary attack. Stay alert.

TERRORIST ACTIVITIES

I was working at my computer one Tuesday morning several years back when my wife told me that a plane had hit the World Trade Center in New York City. We're an Air Force family, so she assumed that the pilot of the crashed plane had been an amateur. Only a few seconds of watching the TV news clarified for me that we were looking at a terrorist situation. I was a National Guardsman at the time, so I immediately called in to report for duty and say that I was on alert, with my bags packed and ready to go.

People the world over have been dealing with terrorism for decades before 9/11. But that was the day that average Americans perked up and began to understand what the rest of the world has been dealing with. We all witnessed what national security experts envision in their nightmares.

The mindset of a person who pulls off an act of terrorism is different from that of an armed individual who mugs you on the street, or enters a bank intending to steal cash and valuables. Muggers and robbers want to enrich themselves.

Terrorists, in contrast, aim to kill as many people as efficiently as possible, and to leave survivors with a permanent sense of dread. To accomplish that goal, they use any weapons that they can get their hands on. Explosive devices are the easiest to construct, and have the singular advantage of being able to be placed in a location, rigged, and detonated from afar. A small bomb can do a lot of damage.

But anything can be a weapon. Cars and trucks can be weapons when driven into a crowd. A mailed package rigged with explosives can be a weapon. On 9/11, the weapons of choice were four hijacked jet planes. In 2009, a young would-be terrorist aboard a flight from Amsterdam to Detroit tried to set off plastic explosives hidden inside his underwear. He didn't get away with it. But his story should stick with us all. Anything can be a weapon. *Even freaking underwear.*

The terrorist's goal is to change society's behavior. And to do that, they have to target as many innocent people as possible. If a terrorist is "successful," society is permanently altered.

By that definition, the 9/11 terrorists were "successful" in achieving their goals. They changed the way everyone on the planet goes about air travel. Before 9/11, plane hijackings were a common threat. Now that threat has been winnowed down to nothing. Before 9/11, it was common for family members to walk their loved ones right to the boarding gate for their departure flight. Or to wait at that boarding gate for their arrival. That will never again happen. Today, all over the world, passengers alone are permitted in the boarding area, and everyone—even pilots, flight attendants, and airport staffers—undergo strict (admittedly often annoying) personal screening before being allowed access to the gates. Even our shoes are screened, because *another* would-be bomber tried to detonate his shoes on a Paris-to-Miami flight in 2001. I guess we can say that the shoe bomber has also forever altered flight travel. The "new normal" is now so engrained in our psyches that all of us would feel unsafe boarding an aircraft that did not strictly screen passengers.

There's another big point to understand when it comes to terrorism: *The beginning is just the beginning.* Here's what I mean by that. On that infamous Tuesday, two massive jet planes crashed into two buildings, setting off a chain reaction that led to subsequent disasters. I want to call your attention specifically to that chain reaction. An instant after the planes struck, the buildings caught fire, and everyone did what they were supposed to do in a commercial building fire. The office workers evacuated, and the firefighters rushed in to fight the blaze.

But this was no ordinary fire. Heat from the jet fuel created fires so intense that they destabilized the steel beams and trusses holding up the two 110-story structures. The two structures collapsed, ultimately killing 2,606 people. People died aboard the hijacked aircraft, people died in the buildings upon first impact with the two jets, people died when the buildings collapsed, witnesses died on the ground, and first responders died trying to help.

In the U.S. Special Forces, instructors often say that *no plan survives the first bullet.* That's their way of saying that disaster is by nature so catastrophic that it defies planning and defies all contingency measures. On 9/11, emergency room physicians in Boston geared up, expecting to treat victims of the World Trade Center being airlifted to Boston when emergency rooms in New York City reached maximum capacity. They were following *their* protocol, not thinking that this situation was different in all respects. The response of the Boston docs seems naive in retrospect. Those victims never arrived because they were annihilated when the towers fell, a scenario few civilians anticipated.

This "spiral of disaster" is right out of the terrorist playbook. Terrorists love creating a perfect storm. They provide the spark, and let the circumstances do the rest. That's why in any terrorism event, you must force yourself to think the unthinkable: *How can this get worse?*

Some examples:

- **A commercial building or school receives a bomb threat. Authorities evacuate the structure. Terrorists initiate a drive-by shooting, or drive a truck into the panicked crowd, to kill as many of the evacuated bystanders as possible.**

- **A bomb goes off in a concert arena or stadium. First responders arrive to treat the wounded. A second bomb goes off, killing and maiming both survivors and the first responders.**

See what I mean? What you see and experience at first may not be the biggest thing coming. The first threat or bomb may merely be a prelude to a greater, secondary attack. Terrorists are adept at using an initial instigating incident to distract and terrify people, then launch a second attack.

Okay, you're thinking. If anything—even underwear—can be a weapon, how can I possibly fight terrorism? You need to do what I said at the beginning of the book. *Keep your head on a swivel.* Stay on the lookout for anything unusual, and report what you see. As the U.S. Department of Homeland Security says, "If you see something, say something."

One time, long after the events of 9/11 had seeped into my consciousness, I spotted a suitcase sitting alone by itself in the middle of a walkway to the boarding gates at JFK Airport. It was so obvious that people had to walk around it like a rock in the middle of a river. I reported it immediately to an airport worker. "Huh," the worker said, "I don't even know what to do about it."

I sort of understood how the airport worker felt. Since the events of 9/11, Americans have been warned *ad nauseam* to keep an eye out for and to report unattended bags in public places. In airports, especially, this warning is broadcasted over loudspeakers constantly throughout the day. The result being that it is truly rare that a traveler leaves their luggage unattended for long.

But on this particular day, no one was rushing over to claim the bag. The airport worker summoned security, and in the minutes before they arrived, I noticed as many as a hundred people walk by without even pausing to look at the bag. Sure enough, as soon as security arrived, a woman popped out of a nearby store, announcing that the bag was hers. "Well," the security guys said, "Sorry, but can you please come with us and answer a few questions?"

They led her away, and I was left wondering why I ended being the only one who reported the bag. The truth, I fear, is that when we've been warned so many times to do something, we stop thinking of it as critical or important. We mind our own business, when we should be minding everyone's business.

In the horrific Boston Marathon bombings of 2013, the terrorists responsible for the attack placed two backpacks containing highly destructive pressure-cooker bombs loaded with shrapnel about 200 yards (183 meters) apart along the marathon route. No one spotted or reported them. With so many people around, it must have easy to assume that the backpacks belonged to someone. They didn't.

When I first learned to drive as a kid, I learned that a safe driver should always be on the lookout for trouble. You never know if another driver is going to do something unsafe—due to their own inattention or for a reason beyond their control. That's why, when you're driving, it's always smart to think ahead. "If a driver on my left or right came into my lane," you might ask yourself, "what's my escape route?"

That's almost how you have to think with respect to terrorism.

- **Keep your head on a swivel. Any time you are in a crowded place—an arena, a concert hall—do not chill out and enjoy yourself until you have scoped out the nearest exits. Don't be paranoid—have fun but keep part of your brain aware. Often the nearest exits are not the way you entered. Terrorists love crowds because they are assured of the biggest bang for their buck. The more people pack into a space, the greater the risk of danger. When you are moving on foot in a mass of people, keep your eyes on what's going on around you. Especially what's going on behind you. Most of us forget to "check six"—which is military parlance for what's going on directly behind you.**

- **If you're walking on a sidewalk in the vicinity of traffic, you should be just as mindful of the vehicular traffic as you are of the foot traffic. It takes nothing for a larger vehicle to jump a curb. Ask yourself a version of the question I told you to ask your kids back in the section about fires: "If a car came at us right now, what could I do to avoid it?"**

- Cultivate a sense of what's normal, and when something in an environment doesn't fit that criterion, alert others. Learn to do this, even if it makes you feel uncomfortable. In 2010, street vendors hawking their wares on a street in New York spotted a parked SUV that had smoke coming out of an open window. They summoned the police, who discovered a massive bomb in the vehicle, equipped with 250 pounds (113 kilograms) of fertilizer, 60 gallons of propane, and 10 gallons of gasoline, among a ton of other flammable substances. The vendors who called the cops prevented widespread destruction and loss of life in the Times Square area.

- Should you ever receive, at work or at home, a mailed package that you are not expecting, that has overwrought packing tape, bits of wire, or even oil spots, treat it as a suspicious package and notify the authorities.

- When a bomb goes off, if you're not in the area, you will still feel the resulting sound wave. It may feel like an earthquake, with the ground shaking under your feet. Do not wait around expecting instructions or an explanation from authorities. You are the only authority you need to get moving. Head for the exits, and as soon as you can, separate from the larger crowd and strike out on your own. In my work I routinely see news images of citizens standing around after a disaster, holding up their phones to capture footage of the incident. Instagram and Twitter are not worth losing your life over. Once you have extricated yourself from a building, arena, or barricaded area, get out of the area entirely.

- The first people killed by bombs are those in the immediate vicinity. First, they are "hit" by the shock wave of the blast, which can harm internal organs. Next, they are struck by shrapnel and wounded by blunt force trauma. If you escape unscathed, and are moved to assist, apply direct pressure to bleeding wounds and use any article of clothing you can to fashion makeshift tourniquets. (See pages 29–30.)

- If you are all out of options, and in physical proximity to a bomber or terrorist who has tipped their hand, don't waste time attempting to bribe or plead with the person. Remember: They're not moved by personal enrichment or appeals for mercy. Instead, use whatever weapons you have available to fight as if your life depends on it. You have nothing to lose—you may well die anyway—but your actions may well impede the escalation of the attack. Both the shoe bomber and the underwear bomber were foiled by fellow passengers and flight attendants, who took them into custody. On 9/11, the plans of the four terrorists aboard United Flight 93 were foiled by passengers who fought to regain control of the plane and divert it from its intended descent and attack upon Washington, DC. The forty heroic people aboard the flight died in the resulting crash in a Pennsylvania field, but no one was harmed on the ground.

ELEVATOR ACCIDENTS

E very year a minuscule number of people—about thirty—die in elevator accidents in the United States. That number is so small as to be statistically trivial. (Notice I did not say statistically *insignificant*. Those are still real human beings, who leave behind grieving, bewildered families.) Still, your chance of dying this way is 1-in-10 million, a probability that shouldn't keep you up at night. At the same time, every year in the United States *17,000 people* are seriously injured in elevator and escalator accidents.

That's the bad news. People hurt themselves on these conveyances in the most awkward ways imaginable. They trip, they fall, they lose fingers and toes, mostly because they never took the time to prepare themselves for what *could* happen.

The scariest thing about elevators is the fear that the cable holding the car up is going to *snap* and send you falling seventy-five stories to a gruesome death. Let me assure you that this is virtually impossible. Elevators have not one but *multiple* cables holding them up. So, if one snaps, there's a dozen more in place to keep the car from making a horrific plunge. Moreover, there are *multiple* brakes in place to slow and stop a falling elevator.

So how do people die? Most often it's because they panic. Occasionally, elevators stop working for minor mechanical reasons. Or they cease operating in a blackout. (See Blackouts, page 140.) Nobody likes being trapped in an enclosed space, so after a short period of confinement, people may be moved to pry open the doors.

Don't be one of those people.

You never know when the power will return and the elevator will begin traveling again. That's a great way to lose a limb or be smooshed between the elevator compartment and the inside of the shaft. That's exactly how people die in these situations.

In the movies they always show you that it's a piece of cake to escape an elevator: You flip open the hatch at the top of the car and climb onto a waiting ladder. But in real life, that hatch is bolted shut from the *outside,* and there is no ladder.

The smart thing to do in a stalled elevator is remain calm, talk to your fellow passengers, and occasionally yell for help and bang on the doors. Use a mobile phone to call for help. *Don't* jump up and down. That throws off the elevator's delicate balancing system and can jam it further. Instead, keep cool until help arrives and you'll emerge in one piece.

BIOTERRORISM

Bioweapons have been with us for thousands of years. We just didn't call them that. As long as humans have waged war against each other, they've tried to kill each other by any means necessary. And in the days when humans couldn't cook up weapons of mass destruction in laboratories, they naturally turned their attention to pernicious natural pathogens, devising the cleverest ways to deliver the lethal payload to their enemies.

Sometimes the deadly weapon was hidden inside a beautiful container. The legend of the Ark of the Covenant can be read as an early example of bio-warfare. When the Ark fell into the hands of the Philistines, they mysteriously began to die from an affliction of buboes—swollen lymph glands that are the mark of bubonic plague. Another example dates to AD 165, when the ancient Romans laid waste to Babylon. One of the Roman soldiers dashed into a temple of the enemy, beheld a gorgeous golden chest, and used his sword to crack it open. For the next 15 years, Rome suffered a terrible "plague" that modern historians believe was smallpox.

In the absence of a Pandora's Box, people got very, very creative. Mongols who attacked an Italian fort on the Black Sea in AD 1346 simply catapulted the corpses of their own dead into the Genovese outpost. Not just any corpses, mind you. They were, in fact, soldiers who had died of the bubonic plague.

I'm a junkie for this kind of military history because, as I mentioned when we were talking about pandemics, *what's old is new*. After I became an ER doc, I became interested in how doctors could spot the effects of bioterrorism before they were noticed by law enforcement or government authorities. And I have since become an instructor to civilian medical providers and to NATO Special Forces on the topic of bioweapons. Biodefense experts obsess about this stuff constantly. They primarily worry about three biggies—a man-made flu strain, anthrax, and smallpox. Influenza and smallpox are both diseases caused by viruses. Anthrax is very serious disease caused by a bacterium often afflicting sheep and cattle.

Modern bioterrorism is far more sophisticated than flinging a dead dude over a castle wall, but the primary logic is the same. Modern scientists (or terrorists) take a natural pathogen, improve it for dispersal, and militarize the delivery system until it is a far better killer than whatever nature came up with in the first place.

Bioweapons need human cells to spread, or at the very least, to perform their job. The ideal bioweapon is a disease that is easy to catch, but that doesn't kill its victims immediately. Ideally, you want victims to live long enough to pass the germs on to *tons* of other people before anyone realizes what is happening.

By that criteria, smallpox (*Variola* virus) is a perfect bioweapon. A vaccine is the only way to prevent it. There's no cure for it once you've got it. It's easy to pass from one human to another. A smallpox victim remains contagious from the moment the first pustule forms until the very last scab falls off their pockmarked body. And we think the disease kills nearly 30 percent of the people it infects. Ebola virus is not a good candidate for a weapon; it's very deadly, killing about 50 percent of its victims on average, but it's really hard to catch.

If you manipulate spores of anthrax (*Bacillus anthracis*), you can turn them into an aerosolized weapon that kills 50 to 80 percent of victims who inhale it. But the expertise required to do that is rare. Even if you could pull it off, anthrax still isn't a perfect weapon, because victims are not contagious. A committed terrorist would have to figure out a way to expose *lots* of people to the spores, either at one time or individually. They could mail tainted packages to a large number of people (which has happened). Or they could dump anthrax-laced powder off the back of a subway or train and hope that it infects lots of people. On paper, measles (*Morbillivirus*) is a perfect candidate. It's very deadly and stays airborne for hours, lying in wait for people to catch it. But it's really not a practical weapon because billions of the world's citizens have been immunized against it.

Already, you can see how tricky and complicated it would be to mount a serious bio-attack on a large population. That complexity is actually helpful in tracing the origins of an attack. Outbreaks of smallpox, and large outbreaks of anthrax, for example, can *only* be acts of terrorism. Here's why.

Yes, anthrax still exists naturally in certain parts of the world. The bacterium dwells inside large spores whose shells are notoriously hard to break. Spores can last decades. There are at least three islands in the world that are or were contaminated with anthrax because that's where scientists in the employ of various governments experimented with anthrax in the past. Innocent, naturally occurring anthrax cases are rare and isolated—about 2,000 annually worldwide. So if we see a huge cluster in anthrax cases, we have to assume the spores came from a lab. The disease can be treated fairly well with antibiotics, so on paper anthrax feels more like a "panic" weapon, analogous to dirty bombs. Assuming we can get enough antibiotics to treat a huge population, anthrax isn't going to be able to kill a lot of people—but it will scare them silly.

Smallpox, on the other hand, was completely eradicated from the planet in the 1970s. That disease today exists only in labs in the United States and theoretically Russia. At one point during the Cold War, the United States and the former Soviet Union behaved like two kids on a playground:

"You destroy yours!"

"No! Destroy yours *first!*"

Instead of destroying the last relics of the disease, the United States stockpiled enough of the pathogen to be able to create vaccines in case of a bio-attack. We can only assume the Soviet Union did likewise. When the Soviet Union dissolved, its stockpile was "officially" transferred to a lab in Russia. But in the ensuing political upheaval, those killer bugs could have been sold, stolen, or dispersed cooperatively anywhere in Eastern Europe or the Middle East. And to be perfectly frank, the United States messed up too. In 2014, scientists discovered several vials of smallpox in the back of a government lab freezer in Maryland that hadn't been cleaned out since the 1950s—about 70 years ago! The vials were quickly collected by the Centers for Disease Control and Prevention and spirited away into lockdown.

All we really know is that the world has not seen a case of smallpox since the 1970s, but we must always remain vigilant. When I was stationed in Iraq in 2004, I received anthrax and smallpox vaccines because U.S. troops were prepared for the possibility that Saddam Hussein had those germs. (He did not.)

Summing up: Bioweapons are hard to select, hard to acquire, hard to manipulate, and hard to deliver. They are also a double-edged sword. They can easily backfire and kill not only the intended target, but their creators as well.

That's why, as the world saw only days after 9/11 when letters containing aerosolized anthrax were sent to U.S. journalists and congresspeople, limited bioterrorism "scares" are more likely in the future than actual outbreaks.

Five people died in the 2001 anthrax attacks, but I'd argue that the still-unknown terrorists who unleashed the pathogens were gleefully satisfied by

the ensuing panic. They definitely altered the way Washington, DC, operates. One cannot easily mail anything to a U.S. government official any more.

One of the few authentic, *sizable* bio-attacks that occurred in the United States was perpetrated in 1984 by members of a cult whose goal was to suppress the local vote in their county in northern Oregon, so their own people might be voted into office.

On a trial run, operatives visited several restaurants and dumped bags of cultured *salmonella* into the salad dressing and ingredients at various buffets, and in the salsa at salsa bars. The result: 750 people took sick, many were hospitalized, but no one died. The cluster of cases was so obvious that an immediate investigation outed the conspirators, who were unable to follow through with their plot to disrupt the November elections.

That case actually gives me hope. Yes, local citizens were terrified, and perfectly innocent restaurants lost revenue and were nearly ruined, but the case was nevertheless a good example of how community action can combat bioterrorism. The sooner the cases are reported and analyzed, the faster medical teams and law enforcement can spring into action.

The catch with bioterrorism—the thing that scares a lot of really smart people—is preparedness. Are the world's governments really prepared to combat a major bio threat? Many people say no. If five people could die in 2001 from an anthrax attack on U.S. soil, you have to wonder whether that government or any other will be nimble enough in an emergency to treat thousands of people if they are exposed. That's why biodefense experts are tasked with thinking of all the ways individuals can protect themselves. For all the rhetoric and theoretical planning over the years, I would argue that we were still not adequately prepared for the most recent pandemic.

Surviving Bioterrorism

- Ensure that your home HVAC system includes an air filter with a Minimum Efficiency Reporting Value (MERV) rating between 7 and 13. In medical labs, they use air filters of higher quality (aka HEPA), which are capable of filtering out even smaller micro-particles, but those filters are not practical for home use as they impede air flow and could damage your HVAC equipment.

- Anthrax spores are large enough that an N95 mask will block them. Technically, flu viruses and smallpox are small enough to pass through N95 masks, but those contagions don't travel by themselves. Both travel via respiratory droplets which are 5+ microns in size. N95 masks filter down to 0.3 microns, so those masks protect against those contagions. But one caveat: Masks do need to fit well to work the best. If you have a beard, you're probably not going to get the best seal possible.

- To protect yourself from an engineered flu strain, you'd follow the same protocol as you did during the COVID-19 situation. Masks, gloves, hand sanitizers, frequent hand-washing, and social distancing are the key. This is also a bad time to be an anti-vaxxer. The more you get an annual flu vaccine, the more your body will build up an immunity to flu strains. It won't be completely protective in a novel flu situation, but it will be better than nothing. You also want to increase your own health and fitness. Normally, healthy people die in the fewest numbers.

- Here's the good news: Scientists absolutely know how to manufacture smallpox and anthrax vaccines. The bad news is, not everyone is a good candidate for a smallpox vaccine. For example, eczema sufferers can die from smallpox vaccine.

- More bad news: You can't vaccinate 330 million people at once. Getting enough vaccine doses to treat a large population will probably take time. No one has ever attempted fast, large-scale treatments for these diseases in history. In theory, most nations can do it. Practically, the rollout is likely to be haphazard and chaotic. You'd treat huge numbers of people according to a protocol called "ring vaccinations." You inoculate people living in a major city, and everyone in the radius of 100 miles (161 kilometers) outside that city. Slowly, you begin increasing the size of the ring, or radius. If you live pretty far out, the assumption is you have not yet been exposed. But none of us wants to be living on the outermost ring.

- If you have been in a workplace that has received a bioterrorism threat, be prepared to disrobe and decontaminate immediately. Medical teams and law enforcement will most likely be on-site to assist with this and provide facilities. Decontamination is critical. As we saw, anthrax is particularly long-lived. What good is treating the first victims if a building or some other public place remains contaminated? After the 2001 attack, Washington, DC, offices were fumigated and sprayed down with a special foam designed to destroy spore shells.

- If you've somehow managed to return home before you learn of a biothreat, bag your shoes, clothing, and any office items such as your briefcase, and shower immediately. Report your whereabouts to your local public health department, and make a list of all stops you made on your way home. Wear a face mask and self-isolate from your family until you are seen by a physician.

- Not all bioterrorism attacks are announced or claimed by their perpetrators. Be curious. Develop your own "index of suspicion." The truth is, we are all sentinels. We are all canaries in the coal mine. Flus and colds are typical in the winter season, but should you ever begin to notice that several people in your family, neighborhood, or workplace are taking sick with the same extreme or unusual symptoms, speak up immediately. Don't assume that local physicians have noticed and reported the trend to local authorities. Call your local public health department immediately.

- Alarming symptoms include respiratory distress, fever, chills, diarrhea, and *bloody* diarrhea, which indicates that the intestinal tract has been harmed.

- If you are traveling abroad when exposed to a bioterrorism attack, contact your nation's embassy prior to leaving the country. Depending on the situation, you may be asked to self-quarantine prior to your flight, or upon arrival. People traveling to certain countries may require specific inoculations. These may include vaccines for hepatitis, yellow fever, typhoid, and Tdap (tetanus, diphtheria, pertussis). You'll want a prescription for anti-malarials if that disease is a threat where you're traveling. Always travel with medical evacuation and repatriation insurance. Without insurance, your out-of-pocket expenses to be transported home safely on a flight with on-board medical care and providers could easily be a minimum of $20,000.

- Alarming symptoms do not have to be bioterrorism to be deadly. Fifty years ago, foodborne disease outbreaks clustered in one geographic region and were easily spotted. Today, food travels all over the world before it ends up on your plate. From *Listeria monocytogenes* contaminating cantaloupes, deli meat, cheese, salads, or ice cream to *Escherichia coli* in ground beef, nut butters, bagged lettuce, or fast-food burgers, the medical literature is filled with cases of tainted food. And physicians are constantly learning more about the proper way to treat exposure to these ailments. For example, different strains of *E. coli* respond differently to treatment. One strain (O157:H7) releases its toxin when it dies, so doctors learned *not* to administer antibiotics to these patients because doing so would actually make matters worse.

CHEMICAL AGENTS

n 1978, a Bulgarian writer who had defected to the United Kingdom was waiting at a bus stop on Waterloo Bridge overlooking the Thames when he felt something akin to a bee sting on his right thigh. Looking around for the presumed insect, he instead saw a man rushing away with an umbrella in his hand. The writer fell ill, and doctors and investigators later pieced together what happened. His assassin had used a specially rigged umbrella "gun" to fire a pellet containing ricin into his body. Ricin is a highly toxic chemical derived from castor bean oil. The writer—who was highly critical of the Bulgarian government, which at the time was part of the Soviet bloc— was dead in 4 days, and the bizarre story of his umbrella assassination has become the stuff of spy novels and films.

In 2018, a man and woman in Amesbury, a city near Stonehenge in the United Kingdom, discovered a discarded, half-used perfume bottle in a park trash bin, and sprayed themselves with its contents to see if they liked the fragrance. They instantly fell ill. The man survived, but the woman died 8 days later, killed by a deadly nerve agent developed by the Russians. As you might suspect, the bottle they found in the park contained not perfume but enough poison to kill thousands of people. The bottle had been used to

poison the front door handles at the home of two people living in Salisbury, only 8 miles (12.9 kilometers) away, and then later discarded by the assassins. The two intended targets, a father and daughter, both recovered from their attempted assassinations. The father just happened to be a Russian double agent who was working with the British.

A year earlier, in 2017, the half-brother of Kim Jong-un, the leader of North Korea, was attacked in the Kuala Lumpur airport by two women who strangely smeared two different liquids on his face, one right after the other. Alone, the two substances were completely harmless, but when combined, the compounds formed a deadly nerve agent, VX. The target, Kim Jong-nam, died en route to the hospital.

I told you: I'm a junkie for this stuff, and I could spend all day regaling you with tales of how humans have tried to destroy each other through the centuries with chemical weapons. Centuries ago, before the invention of modern chemistry, enemy combatants used whatever toxins they could find in nature to do each other in. They used snake venom, stinging beetles, or even the venom of poisonous frogs. In AD 198, an Iraqi outpost defended itself against Roman invasion by flinging clay pots over their walls containing live scorpions.

Today deadly toxins are more likely to be concocted in a lab than drawn from nature. Modern chemical weapons are usually organophosphates that chiefly work by attacking the central nervous system. The telltale sign of exposure to a nerve agent is the contraction of the victim's pupils down to the size of tiny pinpricks. (The pupils are the black part of the eye.) Next, victims start spewing a ton of fluids—foaming saliva, tears, mucus, you name it. They experience stomach seizures and pain. Untreated, victims go rigid, lose control of all bodily functions, asphyxiate, and code right into cardiac arrest. By the way, I should note that in the ER I have seen patients who have experienced all of these symptoms who were *not* victims of terrorist chemical attacks. In fact, they were farmers who had been exposed to pesticides, which are made from the same class of chemicals.

Antidotes to these chemicals do exist, and treatment is definitely possible, but it's all a matter of getting to victims in a timely manner, and disinfecting them as quickly as possible. If mustard gas is the culprit, you'll see every visible part of the victim's body erupting into blisters. What you can't see is that their internal organs and respiratory passages are also blistering. Chlorine gas, which was once used on the populace in Syria, is a common chemical, but the weaponized version behaves like mustard gas. It blisters skin, blinds eyes, scars lungs, and compromises one's ability to breathe.

Let's be frank: It's really hard to protect yourself from such insidious chemicals. You almost have to have a sixth sense about what's going on, and act fast. Quick decontamination and copious irrigation with water is the only way to get this stuff off the human body. On the upside, these chemicals are water-soluble, so they'll break down with soap and water. But their very solubility also means that they can easily penetrate human skin and tissue.

Chemical agents have been used effectively in wars because the toxin, usually prepared as a gas, can be sealed safely in a container, then fired and dispersed upon enemy combatants or civilians. But gas is not the only preferred method of delivery. In the 1995 attack on the Tokyo subway, terrorists dropped bags of liquid sarin to the floors of the subway cars, and punctured them with the tips of their umbrellas. (I know: What is it with umbrellas and chemical attacks?)

The terrorists immediately disembarked from the trains; the liquid sarin evaporated into a gas, infiltrating the subway cars, platforms, and tunnels. By its very nature, though, sarin quickly dissipates on its own. Despite the military-like precision of the cultists who carried out the attack, this particular threat resulted in 12 deaths, which is remarkably low considering what Tokyo's morning rush hours are like.

Surviving a Chemical Attack

- If you are present during an attack, or suspect an attack is taking place, you have three options, depending on your state of mind, physical agility, and circumstances. You can alert others and law enforcement authorities as quickly as possible. You can attack or attempt to intimidate the perpetrators. (In some cases, terrorists have abandoned their plans entirely because they got cold feet.) Or you can cover your face with some form of fabric and run out of the vicinity. Most of these attacks happen in public spaces, which are luckily equipped with abundant signage. Take the quickest exit leading to fresh air.

- Reality check time: Bear in mind that covering your face is only a short-term solution because nerve agents attack all mucous membranes, including those in your eyes and nose. Even if you have covered your nose and mouth, you can't very well run with your eyes covered. And in any case, your hair, hands, clothing, and whatever you used to cover your face will likely be exposed.

- If you were nearby or ran from an attack, you must decontaminate yourself as quickly as possible. This is no time for modesty. Remember Meryl Streep in *Silkwood*? When emergency responders arrive, it's quite likely that they will demand that you strip down to be washed in an open area with fire hoses. If you are alone, don't waste time looking for soap. Get your clothes off and start dousing yourself with water.

- Skin is designed to slough off and rid itself of contaminants. But your couch isn't. Chemical weapons can be unusually persistent when they seep into the surface of common household or office items. If you survive an attack, but suspect that your home, vehicle, or other possessions have been compromised, you must be prepared to go the distance. Small possessions can easily be bagged and disposed of at a hazardous waste site. But homes and vehicles will require professional cleaning—if that is even possible. The Salisbury, UK, home of the father and daughter targeted in the chemical attack I mentioned earlier was enshrouded in scaffolding, had its roof removed, and was treated like a hazmat site for weeks while workers performed deep cleaning with caustic chemicals. One of the British police officers who investigated the attempted assassination was exposed even though he wore gloves and a protective suit at the scene of the crime. He later unknowingly contaminated his own home; he and his family lost everything to the disaster.

- If you come upon a crowd that has obviously evacuated a confined space such as an office building or subway, stay the hell away from the crowd. If the attack has not yet made the news and you have not been glued to your phone, you have no clue what you might be walking into. Now is not the time to stop and chat up people about what's happening. Keep your distance. One case study after the Tokyo attack found that emergency rooms were inundated with people who were not on the subway at the time of the attack, but who nevertheless wanted to get checked out because they were afraid that they might have been exposed. These so-called "worried well" cases clogged hospitals and did no one any favors. If you feel fine, stay out of the danger zone and let first responders and hospitals do their jobs.

DIRTY BOMBS

What have I been saying all along? If there's one underlying goal of terrorism, it's to strike fear into the hearts of innocent people. A "dirty" bomb is a crude weapon in which some radioactive substance is packed into a traditional bomb along with ordnance and shrapnel. When the bomb goes off, it makes the same splash that other homemade bombs do—killing innocent people and destroying property as shrapnel flies everywhere. But the dirty bomb's power over the human psyche lies in its radioactive payload.

In a nuclear weapon, the radioactive core performs a critical function. It's the fuel that feeds a massive blast capable of leveling cities. Nuclear fission is not happening inside a dirty bomb. It's just a conventional bomb that blows up cars, windows, the sides of buildings, and those people unlucky enough to be in the blast zone. In a coldly academic sense, that's really the extent of the damage. Your chances of dying from a dirty bomb if you are not in the immediate vicinity are near zero.

But that's not how it *feels*. People freak out when they hear that radioactive material has been released into the atmosphere. Even if you aced your physics class in high school, chances are you're still hazy on what radioactivity can do to the human body. Everyone knows radioactivity is "bad," but how bad?

If a dirty bomb goes off in a city, everyone is seized by questions that may not be immediately forthcoming. How much radioactivity was it? What type? What was the radius of the blast? How long will the radioactivity last on sidewalks and buildings? What if some got on me? What if I breathed it into my lungs? Is it safe to use that building from now on? Is it safe to walk down that block?

The result? Panic. Fear. Terror. And the terrorist's work is done.

Luckily, it's extremely difficult to get radioactive material. Nations with nuclear capability keep those materials under serious, lethal-force-is-authorized lockdown. That means that if anyone wants to amass radioactive material, they're going to be assembling it in dribs and drabs from elements commonly used in modern industry. For example, radioactive radium-226 was once infamously used to create luminescent paint for dials of wristwatches. It killed the women who were unknowingly exposed to the chemical when they painted those dials. Radium's half-life is 1,600 years, which means that even after all that time, its deadliness has only decreased by half. If you were a determined madman, you could theoretically buy tons of old wristwatches via online auctions and flea markets, and scrape the toxic stuff off the old dials until you assembled a big enough payload. But boy, would it take time.

In the early 1990s, an American teenager was caught attempting to build a nuclear reactor in a shed behind his home in suburban Detroit. He apparently became a tad obsessed while working toward his atomic energy merit badge in the Boy Scouts. The FBI discovered that he collected the radioactive stuff from things like campfire lanterns (which have a little bit of radioactive thorium-232), smoke detectors (americium-241), the glow-in-the-dark night sights on guns (tritium), and yes, old wristwatches. So, not only is it possible to collect material in this fashion, the kid amassed enough of it to build a functional neutron source. The shed in which he worked has since been declared a Superfund hazardous waste site. I don't know if he ever got the merit badge, but there's no doubt he earned it.

Surviving a Dirty Bomb

- The best defense in a dirty bomb situation is to put as much distance as possible between yourself and the radiation. But when a crude, handmade bomb goes off, we're so focused on the very real predicament of those killed or injured in the immediate blast zone that we're not thinking of the possibility of a dirty bomb. Immediately after detonation, bomb sites become magnets for first responders and Good Samaritans who want to help. If you are a survivor and have no skills that can be immediately helpful, clear the area when first responders have arrived, and put as many structures as possible between you and the zone as you leave. You can always give a statement later to law enforcement.

- If you serve in any leadership capacity, you must be thinking two steps ahead: How can this get worse? Could radiation have been released? The area must be tested quickly to determine if that's the case.

- If you're in a building in the vicinity, it may be wise to shut off whatever heating/cooling systems are in use in the current season until you know for certain that radioactive contaminants have not been dispersed into the atmosphere. And keep the windows closed. This may well mean that your structure will become temporarily unsuitable for occupation.

- It's smart to evacuate workplaces or residences in the hot zone. Just do it in a smart way. Anyone who leaves the premises should be moving *away* from the blast area—not toward it. Not for any reason.

- Not all radioactive particles are capable of penetrating human skin. Alpha particles are the best-case scenario. Gamma particles are the worst. Only specialists in radioactive cleanup and protection will be able to help physicians make an assessment on your level of contamination. If you're concerned that you may have been exposed, decontaminate as quickly as possible. Remove and bag all your clothing and possessions. Shower to remove what you can from your skin, hair, and nails.

- Make plans to get tested at a medical facility as soon as possible. In most cases, local authorities will issue instructions about which facility to use and how to dispose of your bagged hazardous materials safely. Phone ahead, be clear that you may have been exposed, and follow any instructions you receive about how, when, and where to enter the facility. I once had a patient who was exposed to copious radioactivity in an industrial accident walk right through the hospital lobby while I and my team were prepping to escort him, and him only, through a designated entrance. Not fun. The entire lobby area had to be evacuated and decontaminated afterward. Keep calm and follow the instructions.

- If you live near a nuclear power plant and are concerned about the release of radioactive materials, you might consider stocking up on potassium iodide supplements. Radiation settles in the human thyroid gland and can cause health problems years later. Potassium iodide blocks absorption of that radiation.

SHOOTERS/ STABBERS

A ctive shooter situations have become a sad fact and a highly politicized part of modern American life. Elsewhere in the world, where would-be assailants have a harder time accessing guns, stabbing rampages are the more likely scenario. In either case, the experts say, your best bet in these situations is to run, hide, fight, and tell others—*in that order.* I have only a couple of slight amendments to these tried-and-true instructions, which I want to delve into, because key components behind the logic often go undiscussed.

Yes, absolutely, at the first sign of trouble, you *run.* You're at an outdoor concert, or you're in a shopping mall, and you hear what sounds like gunshots, you don't waste time. You bug out, *and fast.* You want to put as much distance as possible between yourself and a potential assailant. This makes perfect sense, right? Even the best shooters are less accurate at long distances. And the farther you are from a knife-wielding attacker, the less likely you are to come in contact with the cutting weapons.

This is all fine advice, but look at the great unsaids:

1. **You need to know the position of the attacker in relation to you. Otherwise you could be running straight toward him.**

2. **You need to know the locations of your realistic avenues of escape. You won't know that if you haven't sized up the entire situation from the moment you arrived on the scene.**

When attacks happen in indoor spaces, the biggest mistake people make is trying to retrace their steps and run back to the entrance at which they entered the building. And I get that. Big shopping malls and stadiums are huge, confusing places, and we often train ourselves to get back to where we parked our vehicles or disembarked from public transportation. But in sprawling complexes, your first entry site may be too far to run in an emergency. People become confused, lose their way, and become sitting ducks for assailants, never realizing that there are countless other exits around them that could have led them to safety.

It all comes down to our age-old piece of advice: *Keep your head on a swivel.* Every time you walk into a theater, a shopping mall, a sport stadium, or a concert arena, the first thing you should do is ask yourself: "If something bad happens right now, how would I get out of here?" It may feel like a downer to think this way, but I assure you that once this becomes second nature, you'll mentally file the location of the nearest exits in your brain, and go right back to having fun.

Here's a little trick you can use. If you're sitting in a sizable room among a large group of people, divide the room into imaginary fourths and seek out an escape route in every quarter.

In an actual emergency, if a way out is not obvious, and you're trapped, the next best thing is to hide in a space with a sturdy door, and to lock and barricade that door with many heavy objects—furniture, store merchandise, etc.—as possible. If the door has a window, obscure it and stay out of view. The reason we barricade is twofold. It creates more work for the attacker, and a barricaded door cannot easily be breached with automatic weapons. The attacker has to put muscle behind it, even if he has destroyed the lock or shredded the slab. Keep away from that door and the barricade. The ideal is to be out of sight, out of mind. For this reason, it's often not a good idea to hide with a lot of other people. Although, admittedly, you may not have a choice in the matter. Just remember that a mass of people is easier to spot, and easier to target. You want to make yourself a difficult target.

Assailants of this type are still working from the terrorist playbook. They are forever asking themselves, *What's the best bang for my buck?* They want to kill as many people as possible in the shortest time possible. Unless they are completely deranged, in the back of their minds, they must know or suspect that their window of opportunity is limited. Law enforcement is on the way. Time's ticking. So killers tend to conserve their resources. The sad fact is, most of the time, they don't have to work too hard. Killing unarmed people is frighteningly easy.

Lastly, if you have no other choice, you must find it within yourself to become a rabid dog and fight like hell. Use whatever is handy and easily wielded—chairs, department store clothing racks, hardcover books—to beat the assailant. Bite, gouge eyes, kick—whatever you have to do.

Keep in mind that a stabber is more dangerous the closer you get to him. You want to approach that person with a weapon that can keep him at bay. A person who is holding what we call a long gun, i.e., a rifle or shotgun, is actually vulnerable within a few inches of their person because they don't really have the room to maneuver their weapon into place to take a shot.

Now, please do not misunderstand me. I am not saying your first order of business is to take out a shooter or stabber. Absolutely not. This is highly dangerous, and should only be done if you have no other choice. Animals in nature do what I'm telling you to do. At the first sign of danger, they flee, and fight only to protect their young or when they are cornered. Somewhere in your brain, you are governed by the same instincts. Fighting like hell just might save your life, or it may buy time for other innocents to escape.

In August 2015, on a train bound for Paris from Amsterdam, a man ducked into the bathroom and emerged minutes later armed to the teeth. He had an assault rifle, a ton of magazines, a pistol, a knife, and a small amount of gasoline. He opened fire, but was subdued by seven passengers, all men of different ages, from various walks of life and various nationalities. British, French, and Americans. Every single one of them knew that they could have been killed trying to disarm the gunman. But they leaped into the fray and fought him anyway. One dude, an American serviceman, finally got the gunman in a chokehold and rendered him unconscious. Three people were wounded in the attack, as was the gunman, but not a single person lost their life on that train that day. All seven heroes later received the Legion of Honour, France's highest award, for their bravery.

In November 2019, when a stabber ran amok in London, a number of bystanders fought him and keep him from harming others. They were attending a conference in a 700-year-old historic building on the Thames when an assailant with two knives taped to his hands announced he was going to blow up the building with the "suicide vest" he was wearing. Bystanders used chairs and a fire extinguisher to keep him from penetrating further into the building. My favorite was the dude who grabbed a 5-foot-long (1.5-meter-long) historic narwhal tusk off the wall and used it to keep the attacker at bay. (The building was devoted to the U.K.'s fishing industry.)

The assailant fled the structure and ran onto nearby London Bridge, stabbing five, two fatally, before he was cut down by police. So yes, people died tragically that day, but the number might well have been higher if the stabber had not been cornered and occupied by a handful of heroic civilians.

Tips for Gun or Knife Attacks

- You may have heard that it's best to run from a shooter in a serpentine path, because doing so makes it harder for the shooter to take aim. Forget that. Even soldiers I've trained with don't do that. It takes far too much time to reach shelter or a barricade, and most assailants these days are not actually taking aim at specific individuals. They're using automatic weapons that spray a hail of bullets. The shortest distance between two points is a straight line—take it. Don't worry if you're out of shape. I assure you that in such an emergency, adrenaline is on your side.

- When fleeing, keep your head on a swivel even as you run. You never know if the assailant is flushing people out so that they can be cut down by a second shooter or stabber. Stay alert!

- The fourth bit of advice about shooter/stabber attacks is *Tell others.* A mobile phone in your pocket is both a blessing and a curse. Silence your phone immediately during an active shooter/stabber event so its sounds and lights don't reveal your location. When you have reached a safe place, use your phone to notify law enforcement and/or family members of an ongoing situation. Do not attempt to capture photos or video of the assailant. Avoid social media. Never reveal your hiding location publicly. Do not trust anything you may happen to see about the event on social media. When the larger community learns of an attack, misinformation begins flowing, even from reputable news media. You have no way of knowing if the assailant or his co-conspirators are looking at social media, or using social media themselves to spread disinformation and lure innocents out of hiding. Mobile phone use is a distraction you simply can't afford when your life's on the line. Conserve your battery and keep your head in the game.

- There have been scenarios in which attackers have mingled with fleeing crowds, hoping to escape detection or to launch an attack on law enforcement. In the midst of chaos, police will have no clue who is friend or foe. Do nothing to alarm uniformed officers when they arrive. Keep your hands visible and empty.

- If you need to treat a friend, loved one, or stranger who has been harmed, remember my advice on basic emergency medical skills. Apply direct pressure and be prepared to use articles of clothing to fashion tourniquets. Assault weapons used by many shooters these days are designed to kill, not wound. Each round fired carries a great deal of kinetic energy, which creates a small entry wound, tons of damage inside the body, and a huge exit wound. Similarly, knives are capable of horrific wounds because they can cut far more tissue than a single bullet. Do all you can to help, but realize that what you'll be able to do for victims may be limited.

- It is a tragedy that school districts in some nations must now allocate precious funding to hold regular active shooter drills and teach children how to perform first aid on their classmates. That said, I have found young children to be remarkably resilient in the face of disasters. If you want to talk to your children about shooter/stabber threats, you'd do well to listen more than speak. A simple opener like this should suffice: "Are you worried about someone coming to your school and hurting people? What are your thoughts and feelings about that?" Ask. Get them talking. And listen. That's all you have to do. When they're done, offer advice and reassurance without upsetting yourself as you talk about it. Obviously, the way you prep an 8-year-old for this scenario will be different from the way you speak to an 18-year-old.

- When the chips are down, summon your inner MacGyver. If you are in the assailant's path and cannot get away, use anything to defend yourself. Keys. Umbrellas. Briefcases. Backpacks. Baseball bats. During a 2017 London Bridge attack, terrorists drove a van into a crowd, then ran out, slashing at people with ceramic knives. Bystanders fought them with things like a skateboard, a packing crate, bottles, chairs, and their own fists. One bystander was photographed fleeing from the scene with a pint of beer in his hand. The next day, this triggered the usual jokes in the media about Brits and their beloved pints. Survival experts vindicated the man, saying that in a violent attack a pint glass makes as good a weapon as any. You can fling beer in an assailant's eyes. You can throw the pint at him, or bash him in the face or head with it. And you should, if you are cornered and have no other way to defend yourself. After all, narwhal tusks are not always handy.

CIVIL UNREST

I was young and still undergoing training as a pilot when I got my first whiff of a riot control staple. Our instructors packed me and my fellow pilots into a shipping container and told us to put on our gas masks. They tossed in a tear gas canister and bolted the doors after us. Ensconced as we were in utter darkness, we could still make out the swirl of gas unfurling around us.

This is nothing, I thought. *A piece of cake.*

Then one of the instructors yelled through the door. "Okay, take off the masks."

We did. OMG, I'll never forget what happened. Those eye-stinging, snot-ridden minutes of my life felt like painful hours. What I distinctly remember is how every little break in my skin—even a nick in my face from that morning's shave—burned as viciously as the sensitive membranes in my eyes, nose, and throat. Tear gas is like chopping onions in a hermetically sealed chamber, only ten thousand times worse!

Tear gas is not really a gas. It's a man-made solid or liquid that is deployed as an aerosol in riot control situations to disperse a crowd. Riot police have other weapons in their arsenal as well—everything from water cannons to rubber bullets, flash-bangs, bean bags, and the like. Tools of this sort used to be called "non-lethal" weapons. But any doctor who has treated someone who has lost an eye, spleen, or kidney to a rubber bullet can tell you otherwise. That's why, if you notice, law enforcement agencies have begun referring to these tools as "less-lethal" weapons. They can still kill, just not as readily. For that reason, I strongly urge caution whenever you're in the midst of a large crowd with a potential to become unruly. If you suffer from pre-existing respiratory or cardiovascular conditions, less-lethal weapons may be extremely hazardous to your health.

In most cases, the aftermath of a natural disaster has the effect of bringing communities together. People work together to shelter and feed each other, get utilities restored, and get cities back to normal. But some events have the opposite effect. They're flashpoints that lead to bigger problems. In July 1977, lightning caused a power outage and plunged New York City into darkness. Maybe it was the dicey economic situation at the time, or the politics of the day, or the sweltering summer heat. New Yorkers took to the streets in droves. Looting and vandalism were rampant for 2 days and nights. In the end, 4,500 people were arrested.

That's one form of civil unrest. Another type is when people band together to protest injustice. Even when organizers intend for mass gatherings to be peaceful, it can easily erupt into something else entirely. Why? The simple answer is, *Why not?*

There's an old saying ascribed to grandmothers: "Nothing good happens after midnight." I've found that it doesn't have to be midnight for bad things to start happening. While the "wisdom of crowds" is a documented phenomenon, I don't think anyone can seriously argue that there's a wisdom to mobs. When it's nighttime and people are surrounded by tons of other aggrieved citizens, it's easy for some people to assume that there's little chance of blowback if they act out. *So what if I throw this rock at a store window? So what if I punch this dude in the mouth? Who's gonna know it's me?*

The next thing you know—chaos.

The courage to do something stupid often does not survive sunlight.

Whether you have voluntarily joined a protest, or are simply caught up in crowd situation that suddenly turns violent, you ought to be prepared for what can happen.

Surviving Civil Unrest

- If you're voluntarily choosing to attend a protest that could possibly turn violent, bring along goggles to protect your eyes, copious water to flush your face if you are gassed, and possibly a bike helmet to protect your head. These can all be stowed in a backpack. Leave your mobile phone at home; it can easily be used to track your movements. Yes, I know you're a fine, upstanding citizen and you intend to obey the law. Understand that even if you abide by the law, but a criminal act took place in your vicinity, law enforcement might have grounds to arrest you. If you're visiting a country other than your own, you may not enjoy the same protections as you would at home. Therefore, if you insist on bringing a smartphone, disable anything that can pin down your whereabouts, such as Wi-Fi, location tracking, and cell service. Ideally, the phone should really be on "airport" mode the whole time you're at the event. Resist the temptation to text anyone while there, unless you are using a high-security app. Shut off your photo app's geotagging. Use a numeric password rather than biometric data to unlock the phone. If you must shoot photo or videos, do so from behind a locked screen. That way, if anyone were to confiscate the phone, they'd instantly be locked out and be unable to use any of the device's features to prove your whereabouts at a specific time, or even your attendance.

Bringing a child, senior, or someone who needs special assistance to a protest can be problematic, since you will have to attend to your needs first before you can assist them. However committed you are to the cause, at the first sign of trouble you must be willing to escort your friend, child, or loved one out of the danger zone as quickly as possible. Decide before you go if you are capable of making good on this implicit promise of protection.

The "CS" tear gas (*O-chlorobenzylidene malonitrile*) used by most police forces has been banned from use in declared warfare under the Chemical Weapons Convention Treaty, but many nations use it on civilians. Don't count on the use of tear gas being announced in advance. If you see a cloud of gas-like particles wafting through the crowd, run in the opposite direction as quickly as possible. Pull on the goggles while moving. Pull on a hood if you have it to protect your hair and head. Cover any bare skin. Unless the goggles have an exceptionally tight fit, they will not be effective at locking out tear gas. And even if they were, your mouth and nose would still be vulnerable. Putting them on will, however, protect your eyes in case rubber bullets are fired. If the gas hits you, you will probably not be able to leave the scene easily because CS molecules seek out and bind to moisture— sweat, mucus, hair oil—causing coughing, sneezing, and incredible pain. There have been cases where tear gas has been blown by the wind into neighborhoods and locations that were not the intended target. If you can, keep moving until you know for certain that you are out of the danger zone. Once you are completely out of the area where the tear gas has been dispersed, turn so your body is facing into the wind. Any lingering gas will be blown away from you. Pour as much water onto your eyes and face as possible, allowing it to run away from you and onto the ground. Like it or not, your day is over, because you really ought to get home and shower to remove any lingering CS molecules from your body. Launder the clothes you were wearing as well.

Authorities may also deploy OC (*oleoresin capsicum*), which is a natural product made from hot peppers and can sting just as badly as tear gas. It's launched at people either in the form of an exploding ball (sort of like a paintball) or from a hand-held pepper spray can. The trouble is, in the heat of the moment, this agent won't look very different from tear gas. Resist the temptation to rub your eyes or nose, since this will only make the stinging worse. Since OC is oil-based, it will not easily wash off with water. Some self-defense companies sell antidote sprays that can be used in the moment. If you have not come prepared, you will need to use soap, shampoo, or a mild detergent to wash the substance from your face and body.

Common crowd-control weapons are projectiles fired from special guns. The projectiles can be anything from sponge-tipped bullets or bean bags to bullets made from rubber, plastic, foam, or even wood. Every law enforcement agency has rules about how close they can be standing to a civilian and still fire these weapons. Being shot with a bean bag doesn't sound very dangerous, but those bags are filled with lead. Rubber bullets sound like they'd bounce right off you, but it's a very stiff form of rubber, and it can easily damage internal organs and eyeballs, or crack a rib. The human temple is the thinnest part of the skull; a direct hit could easily fracture the skull. Police tend to fire projectiles at the ground, so they lose some impact as they ricochet up and hit civilians. Your only real defense from these types of weapons is distance and thick, padded clothing.

Flash-bangs and stun grenades might best be described as "disorientation" bombs. They're thrown or tossed in the vicinity of crowds, and explode with ferocious light and sound. (These devices typically use magnesium—the same substance in campsite fire starters [pages 63–65]—which burns extremely hot.) All you want to do when one of these goes off in your presence is run. The flash of light is blinding, and the sound can rupture eardrums. Noise reduction earmuffs like the ones used on construction sites or by landscapers can help, but again, you will get little warning when disorientation devices will be employed.

Aerosol weapons are used for riot control, but they are also available for self-defense in small spray canisters. These days, OC pepper spray is the one you're most likely to be able to buy and use for personal protection. Every state and nation has its own laws governing how much you're allowed to carry, and how you can transport it. For example, 4-ounce canisters of OC spray are permitted in checked luggage by the TSA, but not as a carry-on. And any spray that contains more than 2 percent of CS or CN (*chloroacetophenone*) compounds is banned on flights. The firms that sell pepper spray tend to sell test canisters, so you can get a sense of how to operate their special caps and dispensers in an emergency. They temporarily incapacitate attackers, giving you a chance to flee and call for assistance.

CYBERATTACKS

Years ago, I noticed with some amusement that my daughter had a tiny Post-It stickie thing taped over the camera at the top of her laptop computer screen. "You mind telling me why you do that?" I asked her. She launched into a story that turned my hair green. While she was attending college, she hung out with a gaggle of computer science majors, all of whom blocked their camera in the same manner. One day, one of them showed her just how easy it was to access and control the camera of another student's computer. They could actually spy on another student without that person's knowledge, and all *without* turning on the little light that usually switches on when the camera is in use. From that moment on, my daughter's laptop has never been without a stickie, unless she removes it for face-to-face calls that she herself initiates. "Dad," she told me, "if a bunch of computer geeks were doing it, I figured I should do it too."

The Internet was supposed to be a good thing. A force for change. A way to democratize the world through free and open communication and the sharing of information. But now, not a day goes by without us hearing of hacking events that compromise our precious data and personal information. We live in fear that highly sophisticated thieves will access everything from embarrassing personal photographs stored on our phones to our life savings to the critical functions of our businesses or homes.

The average person has now amassed between 80 and 200 passwords, depending on their line of work. We're warned to change those passwords constantly to protect ourselves. We hear stories of shady operators all over the world who do nothing but busy themselves trying to hack businesses, databases, and election results. At the same time, we're constantly spammed by fake Nigerian princes, Italian nobles, or estate attorneys who generously want to share with us someone's munificent, seven-figure windfall of cash. As I was writing this book, the news was filled with the story of a 17-year-old kid who somehow masterminded a breach at Twitter, the social media company, and managed to gain access to the accounts of major news figures, including, incredibly, the president of the United States.

Those are the cyber-breach stories that make the news in a big way. We've even reached the point where we joke about those Nigerian princes. What we don't usually hear about are the smaller stories that impact ordinary people.

In my field, medicine, we've seen a spate of ransomware attacks on small, rural hospitals. Marauders will break into a hospital system from afar, steal its data, and demand payment for its return. In 2020, a woman died in Germany because the hospital closest to her had been hacked, and she was sent to another hospital 20 miles (32.2 kilometers) away. The delay in treatment was enough to cause her death. It was believed to be the first time a patient died as a result of a cyberattack on a hospital.

Ransomware attacks have also occurred to small school districts as well. These schools and hospitals are vulnerable because, unlike large institutions in major cities, they don't have a dedicated force of IT professionals working day and night to prevent such breaches from happening. Many small businesses around the world are in the same position. On the promise of improved efficiency, they moved their information online, not realizing that the price of this convenience was eternal vigilance. Unfortunately, these types of attacks are migrating now to even larger institutions, as hackers become bolder and more sophisticated.

Good Web Hygiene

- Be vigilant about which websites you visit, and which emails you open on your computer. Yes, your email software is designed to catch spam before it reaches your inbox, and your browser should flag potentially dangerous websites before you visit them. But it does not hurt to train your own internal spam detector to spot red flags. Anytime you receive an email from a person or firm you don't know, exercise caution. The whole point of these so-called phishing scams is to get you to click on a link or open a document that will allow the scammer to steal information. Some go to extraordinary lengths to do this. They design emails that look very much like emails you've received from companies that you deal with on a regular basis, such as Netflix, Amazon, your bank, or major government agencies. The logo of the company may even be embedded in the email and look quite convincing. The message will claim that your account needs attention, that you have an invoice to pay, payment details to update, or that they have noticed strange activity on your account. It's all untrue. I find that these emails typically look a little "off" to the trained eye: You'll spot misspellings, images look fuzzy, the English is slightly awkward, and so on. If you need a convincer, disable your Wi-Fi for a moment, and hover over the sender's email address. You'll see that the sender's address is clearly not from the firm it purports to be. Mark it as junk so you can train your spam filter.

Even if an email does not contain dangerous links or files, the text can be malicious in nature. A stranger offering you vast sums of money is almost always a scam. The old adage holds true: If it's too good to be true, it probably is. Emails claiming to have recorded you without your knowledge are scams, as are ones that offer an old *true* password of yours as evidence that you have been hacked. (News flash: Lots of databases have been breached. Your email and passwords are certainly already among the swiped. That's why we change passwords regularly.) Whether terrifying or hilarious, these types of messages will never be retired from the game because enough people fall for them. Delete these as soon as you get them, preferably unread. Like every piece of direct mail you get from a legitimate retailer, scammers can often detect if an email has been opened, your geographical location, and even the device or software that was used to open it. Don't give them the satisfaction of knowing that they've found another live one. Delete, delete, delete.

How many times has a friend or colleague told you weeks or months after you've sent a particular email that they found it in their spam folder? I'll never understand people who have hundreds of messages languishing in their spam folders. Make it a daily practice to glance at those messages in case there's anything important among the chaff, and then scrap the junk.

Some firms encourage customers or employees to forward them any suspicious emails for further investigation. This is a perfectly wise protocol. But I prefer not to interact with such emails any more than I have to. Follow your employer's protocols for dealing with these. (Some offices prefer that you forward these to a special department that will investigate them.) At home, the ideal should be to consign them to your spam folder without opening them, and then delete them.

- If it troubles you to delete messages unopened, because you fear losing something important, let me gently suggest that most of the legitimate emails we get in a single day—whether personal or professional—are astoundingly trivial. Really. If you delete a legit email, I promise you that the person who sent it will most likely write and ask, "Hey, did you get my email?" It's not the end of the world.

Spam Calls

Spam callers have become slightly easier to deal with over the years, since none of us these days needs to answer a phone call without knowing who is on the other end of the line. To their credit, mobile phone carriers have begun experimenting with technology that labels certain calls as "spam risk" or "known telemarketer." But that still does not reduce the number of annoying calls you may receive at your home, work, or mobile phone. And it does not reduce the number of outright scammers who are attempting to get you to compromise personal information.

- Simply ignoring calls you don't recognize, and sending them to voicemail—whether at home or on your mobile—is a perfectly fine practice. If you can, set your mobile phone to silence unknown callers. If someone legitimately needs to get hold of you, they will leave a message. Most spam callers will not. Understand, however, that following this protocol will be inconvenient at times. For example, if you're expecting a delivery at home and don't have the mobile number of the delivery person in your address book, the call will go to voicemail and you'll need to call the person back. Small price to pay to weed out the junk.

- If you answer a call from a number you don't recognize, you could choose not to say anything at first, and just listen for a few seconds. A human caller will speak up. But a robocaller will not, at least for the first couple of seconds, until the call is routed to a live telemarketer.

- Never give out personal information over the phone unless you are absolutely certain you are speaking to an authorized individual. Never cooperate with callers who say they're phoning from your bank, your insurance company, your cable company, a creditor, etc., unless you initiated the call. You can always hang up and call the number you normally use to contact that firm or institution, and check if the prior call was legitimate.

- Never answer any questions asked of you by any caller you don't know, especially ones where they are obviously trying to get you to say "yes."

- Never click on links, respond to unknown individuals, or participate in "surveys" that have been sent as texts to your phone.

- Coach your children and elderly family members on how to deal with these types of calls as well.

Password Protection

- In my media work, I constantly see news articles in which computer security experts reveal the most commonly hacked passwords. It blows my mind that someone living in the 21st century would still use passwords like 123456, Qwerty, ABC123, or Ilovemom—but people do! From this moment forward, commit to using complex passwords. In general, the longer the password, and the more varied the characters, the stronger. One technique is to start with a phrase that is familiar to you, and simply alter some of the characters with uppercase letters, numbers, and symbols. Barring that, you can install password manager software on all the computers your family uses, and train the crew to start using them religiously. The best managers generate tough, impregnable passwords on the fly, save them to your vault, automatically and safely fill them in on the websites you visit, and update them when you change the password. Most programs allow you to specify the length of the password, and how many special characters it will contain. Occasionally, you will encounter websites that will not permit password software to fill in passwords for you. They'll require you to type in each character, one keystroke at a time.

- If a website offers two-factor authorization (2FA), enable it. This is just one more security check that helps keep your account safe. Each time you log in, after you enter your user ID and password, the website prompts you to enter a six-digit code. Those codes are typically sent as a text to your smartphone. I recommend that you stop sharing your mobile number with websites. It's fairly easy for a hacker to clone your mobile phone number. (If you want to be horrified, Google your mobile number sometime. You'll see that the whole world knows who owns that particular number.) Use an authenticator app instead. Such apps generate one-time six-digit codes, which you enter on the site to complete your login. If you want a stronger layer of security, buy and use a physical security key (i.e., YubiKey) that plugs into your computer or phone and works in tandem with the authenticator app to generate those one-time passcodes. Without the physical key, a hacker has no chance of replicating your code.

Dealing with Companies Online

- We are all charged with protecting our castles. Institutions like banks, mortgage companies, local utilities, or your favorite retail sites make excellent targets for hackers. You can't do very much to protect your information if *they* suffer a breach. But you can stop storing your payment information on your trusted sites. Yes, it's a pain to re-enter payment details every time you want to buy something or make a payment, but I don't think it's wise to maintain "virtual wallets" all over the web. Who knows? The very act of having to enter that data each time might force you to more carefully consider whether you really want to make that purchase.

- Identity theft is real—still. Shred sensitive documents routinely, and "freeze" or "lock" your credit with consumer credit unions to prevent someone from getting a credit card in your name. If you go this route, you will need to renew these preferences periodically with these institutions.

- Cultivate the habit of checking your various financial accounts on a regular basis. Sometimes, hackers will debit a victim's account for a minuscule amount such as ten cents ($0.10) just to see if the connection works, and if they can get away with it. Don't be so quick to dismiss tiny withdrawals as insignificant. Notify your bank immediately and follow up on them. They could very well be a prelude to a more substantive attack.

- Credit card companies have become increasingly sophisticated about noticing unusual activity. So much so, that it might be wise to notify them when you are traveling to a different state or country so they don't prevent a legitimate charge from hitting your card. (This has happened to me several times, as I travel for work. It's always a pain, but I'm grateful for the oversight.) Be sure that the institution has your most current contact information so they can contact you on the road if they have a question about a charge.

Personal Data

- It's not enough any more to back up your computer to a physical hard drive in your home. In case of an emergency such as a fire, you'd lose everything unless you wasted precious minutes unplugging the device and taking it with you. That's time you could spend ensuring the safety of your spouse, children, and pets. Make a plan right now to regularly update your files to a secure, off-site server.

- It's never wise to store embarrassing images or files on your phone. Many smartphones today are linked to your computer. If your phone is compromised, the images or data are ripe for the picking. If you really must retain that stuff, stick it on an old computer that is no longer linked to the web. Enable your phone's screen lock, and use a strong password to secure it. Or better yet, don't take those photos at all!

Loved Ones and Scams

- Given that many seniors live alone, and may have amassed a sizable fortune after years of hard work, they make ideal targets for scammers. Recent studies have shown that seniors are particularly vulnerable to financial scams. Even seniors who are otherwise sharp as a tack are not able to distinguish lies from truth the way they used to. As we age, our brains atrophy, and our brain cells aren't as quick to make connections that were once second nature. Specifically, two important regions of our brains are impacted: the part of the brain that registers when something important is happening around us, and a region that helps us pick up on body language, voice cues, and other hints about the behavior of other people. As a result, we don't question the motives of strangers. We're more trusting, less skeptical, and don't perceive financial risks or losses as necessarily negative.

Consider the classic scam scenario where a grandparent is phoned or texted with news that a grandchild has been kidnapped, and that ransom money must immediately be sent to some account. The scam works because seniors implicitly trust the veracity of the caller/texter. They become alarmed and act without hanging up and calling their son or daughter to verify the story. Often, adults in the prime of life get angry or upset at their elders because of perceived mental slowness. But it's not their fault. This is a known physical situation that will quite likely strike us all in time. So you *must* make a special effort to protect the seniors in your family. You could, for example, work with your loved one's bank to receive alerts any time they make a large purchase via written checks, credit cards, or debit cards. Be aware that in many instances, the ones most likely to exploit a senior are other family members.

Securing Medical Data

- If you or someone you love is implanted with a cardiac defibrillator device (i.e., a pacemaker), you should be aware that certain models require a firmware update to prevent them from being hacked. This security flaw has not caused any disasters yet, but it is theoretically possible. A person gaining virtual access to the device would be able to alter the number of beats per minute that patient's heart can receive, potentially causing serious injury or death.

- If you must communicate with your doctor outside of the office, I advise using your provider's dedicated patient portal rather than trusting your communication to email and text if you cannot phone the office directly. Such portals are usually quite safe, as providers hire firms that stay on top of potential security issues.

AIRPLANE DISASTERS

Bizarre fun fact: Did you know that passengers almost always have trouble undoing their seat belts and evacuating a plane in an emergency? The reason why will blow your mind. When people get scared, they revert to instinct and do the thing that comes most naturally. Airline seat belts unlock in a manner that is slightly different from the ones we use most often. In automobiles, you push *down* on a little red button. On airplanes, you lift *up* on a metal lever. It's such a little thing, but when you're scared, it's really easy to revert back to the action you have been performing most often for decades, and thus lose precious time getting off a damaged aircraft.

I've abandoned burning planes more often than I'd care to remember. In the movies, every plane has an ejection seat, and every pilot in danger just hits the magic button, shoots out of the cockpit, and sails gently out of the sky with a pair of parachutes easing his landing.

That's not how it is in the Air Force. Turns out, not every plane is equipped with an ejection seat. If your jet catches fire, you've got to figure out a way to notify the fire team on the ground, land that 50- to 60-ton behemoth safely, and get the heck out of Dodge before it incinerates.

You may have heard that you're safer in the air than on the ground. That's absolutely true. Over the course of your lifetime, you have a 1 in 112 chance of dying in an automobile accident, but only a 1 in 8,015 chance of dying on a flight. The chances that the plane you're riding on *today* will have an accident is 1 in 1.2 million. Ninety-six percent of people who are involved in an airline accident survive. These statistics are pretty clearly in favor of flying over driving, but most people don't believe it. Not really.

Airplanes scare people because a) they're in the sky and it feels like a long way down, and b) they're not in control. People feel unreasonably safe when they're behind the wheel of a car. They think that because they're in charge, they will always be able to navigate themselves out of whatever crazy scrapes arise. Judging by the number of automobile collisions that occur each calendar year, they're wrong.

But just because you're safer in the sky doesn't mean that you won't occasionally experience a near disaster. The crashes we all hear about in the news are the *catastrophic* ones. What we don't hear about that often are the near misses that are the commercial analogues to the ones I experienced as a pilot. Minutes after takeoff, the plane reveals itself to have some type of engine trouble. The pilot turns back, lands hastily, and the plane perilously skids to a stop—but everyone's okay.

Here's what you need to know before you fly.

- You may have heard that the safest seats on a plane are the ones in the back. Or the front. Or in the middle. Or, heaven help us, clinging to one of the wings. The truth is, every study that's been done on this question has a different outcome. A University of Glasgow study jibes best with my experience, so I'll mention it. What they found suggests that when you're booking your seats, you want to choose ones that are within five rows of an exit. That single decision is probably enough to save your life. Why? If something happens and you're forced to do a land or water evacuation, you want to be close to a way out. It's as simple as that. If you're farther than five rows from an exit, it will take far too long to get out and your chances of survival start to drop.

- I know it's boring as heck to listen to the pre-flight safety demonstration, especially if you fly often. But you need to know the location of the exits on every plane. As the flight attendants run through their spiel, you find those exits and ask yourself my favorite little question: *If something goes wrong, how would I get out of here?* The laminated card in the seat pocket in front of you will spell out the location of *all* the exits, not just the ones that are immediately visible from your seat. It's absolutely true what the flight attendants always say: The nearest exit might be right behind you.

- By the time we find our seats, most of us are exhausted by the hassle of airport security, and just want to relax. That's fine, but don't take your shoes off just yet. Statistically, most accidents happen during takeoffs and landings, 3 minutes after takeoff or 8 minutes before landing. You want to keep your shoes on in both situations in case you have to bug out. If this feels like an inconvenience, just think how inconvenient it would be to flee a plane by running shoeless over metal, glass, or other debris!

Before every flight, attendants tell passengers that in case of a crash landing, they should brace themselves with their feet *flat* on the floor. But in many airplane crashes, passengers suffer broken legs because they instinctively stretched their legs under the seat in front of them. Don't do this. Be ready to brace yourself exactly as indicated on that laminated card.

Compared to the planes I flew back in the day, passenger jets are equipped with tons of life-saving equipment. Emergency lights. Flotation devices. Oxygen masks. Life rafts. And plenty of exits that make it easy to get everyone off in record time. In fact, every commercial plane is designed in such a way that all passengers can leave the cabin in 90 seconds. But that's only if people think smart and act quickly. Once, in the control room of a TV station, I sat next to an air safety expert as we watched some news footage of people fleeing a troubled plane that been forced to crash-land in a field. After we watched the clip, the safety expert asked us to play it again. "Look," he told me, "watch how many people have their backpacks with them." He was right. Nearly everyone fleeing the plane was toting some kind of carry-on item. That's exactly the kind of thing that will get you and others killed. A plane emergency is like a house fire: You must escape with your life and little else. That's the only way everyone will evacuate in 90 seconds!

Here's what I do: When I first take my seat, I pat down my pockets and make sure I have my wallet, keys, phone, and passport somewhere on my person. That's really all I need. I know that if there's an emergency and I have to jettison everything else I've brought with me, I'll be good to go. Those four essentials will ensure that I'm able to get back home and reboot my life.

When pilots sit in their seats, they fasten their seat belts *tight*. They know that in case of turbulence, tighter is better. Tighter means *less jostling and movement*. The most common scary moment on commercial aircraft is when the plane hits what we call "clear air turbulence" (CAT). Basically, you're cruising along in beautiful weather and suddenly you hit an unpredicted pocket of turbulence. *BAM!*—the plane drops several feet. Everyone screams. Everything goes flying—people, luggage, food trolleys. In simulated exercises, the flight dummies who fared worse in those scenarios were the ones with no seat belts on. I know it's a pain to sit for hours with a tight seat belt, but even passengers wearing seat belts fastened loosely do better than those not wearing them at all. Keep your seat belt fastened when in your seat.

- Something goes wrong. Now you're living your worst nightmare. The plane crash-lands. Get up immediately, leave everything behind, and get moving. In NATO Special Forces, I teach that it is always better to *do something*—even if it is wrong—than do nothing at all. The critical action is *movement*. Why? Simply put, movement allows you to gather intelligence. You might get 5 feet (1.5 meters) up the aisle of a plane only to discover that the closest exit in front of you is on fire, submerged in water, or blocked for any other reason, and you need to turn and run back. That's fine. But if you haven't taken those first few steps, you'd still be in your seat. The person who starts moving is the person who overcomes the ensuing panic and reaches the only viable exit while everyone else is fishing for their precious briefcases.

- To radically reduce your chances of dying on a plane, don't fly. Ever. I'm serious. That's how statistics work. If you're not there, it won't happen. The next safest thing to do is to take *fewer* flights. Fewer flights mean fewer takeoffs and landings. One way to drastically reduce those pressure points is to suck up the cost and take direct flights whenever possible.

- In debriefings, crash *survivors* often say something like, "It was the strangest thing. After we crashed, people were just sitting in their seats, staring and not doing anything." It's not hard to understand why. In terrifying situations, people become alarmed, flustered, panicky, and nervous. What they crave most is someone else telling them what to do. People wait for *permission* to leave their seats. Or they think, "I'll wait to see what everyone else does, and then I'll do that." Or they simply freeze. All of these are wrong moves. As my father used to tell me when I was a kid, "Hesitation kills." Force yourself to take action. When smoke and fire are engulfing you, when water's rising, you don't wait to see what other people do. *You* set the example. Make a decision and act on it! Get to that exit, throw the door open, and be the hero I know you can be.

TAKE YOUR PULSE

T here's a little mental trick every young hospital resident is taught to do whenever a patient "codes"—that is, slips into cardiac arrest. Before they touch the patient, doctors are told to "take their own pulse."

We don't mean it literally. It's just a psychological exercise. A reminder, if you will, to nervous young doctors that nobody is going to get saved if the person in charge doesn't calm down and focus.

When doctors stop and pay attention to what is going on inside them, even if it's just for a second, they will probably notice that their own heart is racing. Of course it is! A fellow human being is in bad shape and they might die in minutes ... and ... they need our help ... and ... and ...

Calm down.

This is the greatest message you can possibly learn that will help you survive any disaster.

Calm down.

You can't accomplish anything if you freak out.

The person in trouble needs your help and your *confidence*. Even if that person is you.

I lived out this dynamic all the time as a young pilot. We'd be flying along perfectly and then hit a pocket of turbulence or a storm that moved more swiftly into our path than anticipated. My co-pilot, who was often younger than I was, would begin to freak out.

"What are we gonna do? We're screwed!"

"Hey," I'd pipe up, "I need you to research airports where we can land. And I'm gonna need that info pretty quickly. Can you handle that for me, please?"

Guess what? Almost always, I *knew* where we were going to land. But the other guy didn't. That task kept him busy—focused and calm—so I could fly us out of there.

Once, when we got back safely on the ground, I had one young co-pilot look at me with admiration that was probably unearned. "Man," he said, "I don't know how you did it. You were so calm! We could've died up there."

I'm glad he thought I was calm. The truth is, I was faking it. I was more freaked out than he was. Shoot, anyone in our situation would have been. But I had to stay calm for *him*. So he wouldn't keep freaking out.

There's another good reason to get calm and stay calm. You're a human being, and all members of our species have the gift of brains, if only we can stop the panic raging inside us long enough to use them.

There's a moment I love in the Matt Damon movie, *The Martian*. At the end of the film, after Damon's character has made it back safely alive after being stranded alone on the planet Mars, NASA puts him in charge of teaching survival training to young recruits.

What does he tell the recruits? "At some point," he says, "I promise, at some point every single thing is gonna go south on you, and you'll think: This is it. This is how I end. And you can either accept that ... or you can get to work."

How do you get to work? His advice is not too different from what I or any other survival trainer would tell you.

Calm down.

Trying to solve all the problems at once is too overwhelming. Your friend's bleeding from three different wounds at the same time. You're lost in the wilderness, it's cold, and you're starving. What do you do? You pick one problem and you solve that. Then you solve the next one, and the next one after that. Pretty soon, you reach a point of critical mass, and you have started to swing a horrible set of circumstances in your favor.

And that's when you save the life of someone you love. You find your way. You pull yourself out of a mess. You fashion a tourniquet. You make a fire and stay warm.

Keep calm. Take it one at a time. That's all any of us can do. It's how we stay alive, and how we save the day. It's how we end up sitting in that rocking chair years later, telling anyone that will listen how we single-handedly fought off three bears in a blizzard as a fire raged around us ...

DR. JOHN TORRES

ACKNOWLEDGMENTS

Writing a book is a challenge for anyone, but it helps when you have an amazing team to rely upon. And in this respect I am probably one of the most fortunate people around. My manager, Carol Perry, supported the project from the very beginning and was unwavering in her support for it.

My agent, Yfat Reiss Gendell, and her team at YRG Partners, sat me down one day and got me talking about disasters. Once I started, I couldn't stop. She coaxed every single germ of an idea out of me, and onto the page for the very first time. That two-page outline formed the basis of a book I didn't know I had in me. Yfat is a super agent among super agents, and this book would not exist if she hadn't made it happen.

Sarah Pelz, my editor at Houghton Mifflin Harcourt, pounced on this idea and championed it every step of the way, nurturing it the way only a great editor can, and sharpening its prose and focus. I'm indebted to her entire team, including publisher Deb Brody, Editorial Director Karen Murgolo, Jess Handelman and Brian Moore for creating the perfect package, Andrea DeWerd and Shara Alexander for their promotional expertise and enthusiasm, Marleen Reimer for selling the book around the world, Johanna Baboukis for copyediting, and Mike Olivo for proofreading, Marina Padakis for keeping us on track at every stage of the process, Christina Stambaugh for her careful eye and expert project management, and Emma Peters, who was in on it from the very beginning.

I thank my co-author Joseph D'Agnese, who made the writing of the book as painless and seamless a process as possible. Thanks to our awesome collaboration during an insane time on this planet, we both have a lot more tools in our bug-out bags to play with.

This book would not have happened without the countless experts over the years who patiently answered my questions, showed me the ropes, and shared with me the irreplaceable gift of knowledge. I am a man who is lucky to have experience in three very different fields—the military, medicine, and journalism—all of which have shaped me to the person I am. To my colleagues at NBC, especially the medical unit, you mean the world to me, and I am lucky to be able to do the job I do every day because I know you have my back.

I owe my life to the parents who brought me into this world and taught a boy and young man that if he could fly toward the stars, he'd probably get somewhere close to them. You are with me always.

Forty years ago, I was fortunate to meet the love of my life and start a family together. From Colorado to New York and the world together. Lynda, you have my heart. And I am proud to have two children follow in my footsteps. Michael John and Allee, you are my world. Just remember to pack the carbon monoxide detector, and you'll be fine.

DISASTER CHEAT SHEETS

Your Prospective First-Aid Kit

☐ Analgesics (i.e., acetaminophen, aspirin [adults only], ibuprofen, naproxen, etc.)

☐ Antibiotic ointments (i.e., Bacitracin, Neosporin, etc.)

☐ Antiseptics (spray, lotion, wipes for cleaning wounds)

☐ Bandages (various sizes of adhesive bandages, rolls of gauze, gauze pads for larger wounds)

☐ Cold packs

☐ Elastic bandage (i.e., ACE)

☐ Flexible aluminum splints (i.e., SAM Splint)

☐ Flexible cold pad (i.e., Thermipaq, stored in family freezer)

☐ Hydrocortisone cream (to treat itching, swelling)

☐ Plastic disposable gloves

☐ Scissors

☐ Self-adherent wrap (i.e., Coban)

☐ Thermometer

☐ Tweezers (pointed-tip version)

☐ Waterproof adhesive tape

International Morse Code

A • ▬	N ▬ •	Starting Signal ▬ • ▬ • ▬	
B ▬ • • •	O ▬ ▬ ▬	End of work • • • ▬ • ▬	
C ▬ • ▬ •	P • ▬ ▬ •	Error • • • • • • • •	
D ▬ • •	Q ▬ ▬ • ▬	1 • ▬ ▬ ▬ ▬	. • ▬ • ▬ • ▬
E •	R • ▬ •	2 • • ▬ ▬ ▬	, ▬ ▬ • • ▬ ▬
F • • ▬ •	S • • •	3 • • • ▬ ▬	? • • ▬ ▬ • •
G ▬ ▬ •	T ▬	4 • • • • ▬	' • ▬ ▬ ▬ ▬ •
H • • • •	U • • ▬	5 • • • • •	/ ▬ • • ▬ •
I • •	V • • • ▬	6 ▬ • • • •	: ▬ ▬ ▬ • • •
J • ▬ ▬ ▬	W • ▬ ▬	7 ▬ ▬ • • •	; ▬ • ▬ • ▬ •
K ▬ • ▬	X ▬ • • ▬	8 ▬ ▬ ▬ • •	+ • ▬ • ▬ •
L • ▬ • •	Y ▬ • ▬ ▬	9 ▬ ▬ ▬ ▬ •	- ▬ • • • • ▬
M ▬ ▬	Z ▬ ▬ • •	0 ▬ ▬ ▬ ▬ ▬	= ▬ • • • ▬

STROKE, SEIZURE, HEART ATTACK, CARDIAC ARREST: Which Is Which?

Is the person . . .

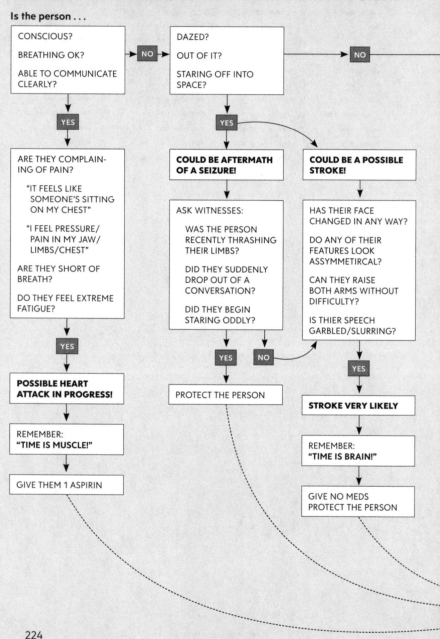

CONSCIOUS?

BREATHING OK?

ABLE TO COMMUNICATE CLEARLY?

NO →

DAZED?

OUT OF IT?

STARING OFF INTO SPACE?

NO →

↓ YES

ARE THEY COMPLAIN- ING OF PAIN?

"IT FEELS LIKE SOMEONE'S SITTING ON MY CHEST"

"I FEEL PRESSURE/ PAIN IN MY JAW/ LIMBS/CHEST"

ARE THEY SHORT OF BREATH?

DO THEY FEEL EXTREME FATIGUE?

↓ YES

POSSIBLE HEART ATTACK IN PROGRESS!

REMEMBER: **"TIME IS MUSCLE!"**

GIVE THEM 1 ASPIRIN

↓ YES

COULD BE AFTERMATH OF A SEIZURE!

ASK WITNESSES:

WAS THE PERSON RECENTLY THRASHING THEIR LIMBS?

DID THEY SUDDENLY DROP OUT OF A CONVERSATION?

DID THEY BEGIN STARING ODDLY?

YES | NO

PROTECT THE PERSON

COULD BE A POSSIBLE STROKE!

HAS THEIR FACE CHANGED IN ANY WAY?

DO ANY OF THEIR FEATURES LOOK ASSYMMETIRCAL?

CAN THEY RAISE BOTH ARMS WITHOUT DIFFICULTY?

IS THIER SPEECH GARBLED/SLURRING?

↓ YES

STROKE VERY LIKELY

REMEMBER: **"TIME IS BRAIN!"**

GIVE NO MEDS PROTECT THE PERSON

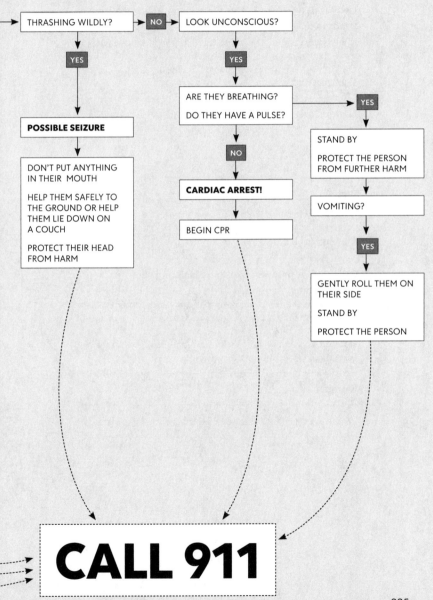

THRASHING WILDLY? → NO → LOOK UNCONSCIOUS?

THRASHING WILDLY? ↓ YES

POSSIBLE SEIZURE

DON'T PUT ANYTHING IN THEIR MOUTH

HELP THEM SAFELY TO THE GROUND OR HELP THEM LIE DOWN ON A COUCH

PROTECT THEIR HEAD FROM HARM

LOOK UNCONSCIOUS? ↓ YES

ARE THEY BREATHING?

DO THEY HAVE A PULSE? → YES

ARE THEY BREATHING? ↓ NO

CARDIAC ARREST!

BEGIN CPR

STAND BY

PROTECT THE PERSON FROM FURTHER HARM

VOMITING? ↓ YES

GENTLY ROLL THEM ON THEIR SIDE

STAND BY

PROTECT THE PERSON

CALL 911

Your Prospective Bug-Out "Bag" Checklist

☐ **Backpack or waterproof bag**
You need something to carry it all in, don't you?

☐ **Dehydrated meals (i.e., brands such as Mountain House, Readywise, Backpacker's Pantry)**
Easily found at outdoor stores and many warehouse clubs, these "just-add-hot-water" meals can last 2 to 5 years on the low end, 25 years on the high end! You'll want enough for 3 to 5 days for each person.

☐ **High-calorie or high-protein meal bars**
Shorter shelf life, but you won't need to stop and boil water.

☐ **Trash bags**

☐ **Water**
Humans need 1 gallon per person a day.

☐ **Can opener**

☐ **Camp stove, pots and pans, and cleaning supplies**
Classic camp stoves and grills (i.e., Coleman) run on small, portable, refillable butane or propane tanks; newer types (i.e., Solo Stove) burn twigs and pine needles, so you're not dependent on fuel access. All are available either at an outdoor store, or via online retailers. Aluminum foil can be handy for certain meal preparations. Just saying.

☐ **Eating utensils**

☐ **Water bottle (i.e., Nalgene or stainless steel)**

☐ **Water filter or filtering system, iodine solution, water purifying tablets**
See how and when to use these on page 75.

☐ **Hand-powered radio**
AM, FM, and NOAA weather are musts. Some brands have built-in flashlights and USB chargers. Some run off solar power; some are hand-cranked.

☐ **Work gloves**

☐ **Knife**
Look for a "combo" knife that has a partially serrated blade.

☐ **Paracord rope**
Easy to cut, tie, or braid for multiple uses.

☐ **Multitool or Swiss Army knife**
Some are equipped with handsaws and whistles, eliminating the need for the next two items.

☐ **Whistle**

☐ **Handsaw**

☐ **Combo folding shovel/pickaxe**
Use to handle hot coals, dig holes, dislodge rocks or debris, etc.

☐ **Combo axe/hammer**
Chop wood, pound in tent stakes.

☐ **Duct tape and Ziploc bags**
Both are insanely useful. Choose an assortment of bag sizes—
pint, quart, or gallon. Keep in mind that freezer and storage
bags are more durable than sandwich-style bags.

☐ **Compass**
The world's most essential navigational tool.

☐ **Wristwatch**
In case your mobile phone or vehicle clock are unavailable. Use to
track heart rates or respiratory rates, or record how long ticks have
been on a person's skin. Choose a tough field or dive watch with
good water resistance. Analog watches can be used as a compass
in a pinch. (For more on the ideal disaster watch, see page 70.)

☐ **Maps**
Of the local area and larger geographical region.

☐ **Waterproof notebook, pens, pencils**

☐ **Weather gauge system (i.e., Tempest, AcuRite, etc.)**

☐ **Important personal documents**
Driver's licenses, medical prescriptions or records, critical
financial and insurance papers, passport, birth certificates, etc.

☐ **Books, decks of cards, games, puzzles**
Anything that will help your family pass the time.

☐ **Laptop computer(s), charger, and backup drives**

☐ **Mobile phone(s)**
Always pack a charger or two. Remember that when reception
is spotty, a text is often the smartest way to get a message to
someone in a hurry, because texts don't require as much of a
signal to transmit as calls do. And texts don't eat up as much
precious battery time.

☐ **Mesh network device (i.e., goTenna)**
Allows you to use your mobile phones to send texts and
your GPS locations over long distances with others in your
party, even if cell towers are not available. The underlying
technology is similar to walkie-talkies.

☐ **Small amount of cash, in small bills**
In case bank ATMs are not available.

☐ **First-aid kit**

☐ **Personal, wearable ID band**
One per person, listing emergency contacts and/or medical
information for those with life-threatening conditions—not to
mention everyone else. Identification becomes critical if your
family becomes separated, an injury prevents communication,
and/or personal documents are lost.

☐ **Prescription meds and over-the-counter drugs**

☐ **Stormproof matches, liquid fuel lighter**

☐ **Fire starter tool**
In case you lose, damage, or finish your matches. See how to
use one on page 64.

☐ **Disposable hand warmers**

☐ **Sewing kit**
Safety pins can be used as fishhooks in a pinch. Needles can
mend clothing or stitch human skin.

☐ **Fishhooks, line, yo-yo reel**
A "yo-yo" fishing reel is simply a plastic reel around which you
wrap your line. It's an inexpensive item that will not take as
much room in your bag as a fishing rod. You can easily locate
videos online demonstrating their use.

☐ **Superglue adhesive**

☐ **Toiletry kit**
Razors, body wipes, tampons, condoms, toilet paper, etc. A portable mirror can be used to signal across long distances in an emergency.

☐ **Soap**
Gentle castile soap (i.e., Dr. Bronner's) will wash nearly everything—your body, clothes, or cookware.

☐ **Hand sanitizer**

☐ **Antibacterial or sanitizing wipes**

☐ **Sunscreen**

☐ **Insect repellent**

☐ **Bear spray/pepper spray**
No guarantee of success, on either human or animal foes. Be sure you have trained in the use of these products, and know how to discharge them in a safe manner.

☐ **Flashlight**
Bring spare batteries. Using the flashlight on your crank radio is hard work.

☐ **Headlamp**
To keep your hands free if you are forced to walk in darkness, whether in a building, city street, or outdoor trail. Battery-powered, solar, or chargeable via USB port.

☐ **Portable jumper cables**
Modern ones, equipped with a battery, can also quickly charge mobile phones.

☐ **Solar panel and charging battery**

To charge your phone when electricity is not available.

☐ **Goggles and face masks**

A set for each person, to keep smoke, dust, and toxic airborne particles out of your eyes, nose, mouth, and lungs.

☐ **Clothing**

Hat, gloves, beanie, socks, fleece, rain poncho, performance layer shirt, underwear, long johns.

You'll want more layers than you think, even in warm seasons. "Performance" outdoor clothing is durable, hand-washable, and dries quickly. Wool will keep you warm even when wet. Cotton—however breathable—will not.

☐ **Tent**

☐ **Sleeping bag/pad**

One per person.

☐ **Tarpaulin**

For under your tent.

☐ **Emergency blanket**

☐ **Mountain bike(s)**

They can go where your car can't or won't, if fuel is no longer available.

☐ **Inflatable pack raft**

In case of floods, but you'll need a hand or battery-powered pump to inflate. If you go this route, make sure you have one large enough for your party, and oars to steer it.

RESOURCES AND WHERE TO BUY GEAR

I n this section, you'll find two types of lists. First, you'll see a series of quick 'n' dirty shopping lists for specific needs mentioned in the book. Below that, I include some online shopping links for the more obscure items, which should save you hours of online searching. Both types of lists are far from comprehensive, but they are good jumping-off points. The products I'm mentioning here—not to mention the brands—should be regarded as *suggestions,* not recommendations. You really need to assess what makes sense for your family. For example, what one family considers a must-have for their home pantry will not interest many other families who don't routinely eat the same items.

FIRE SAFETY SHOPPING LIST

Carbon monoxide detectors

Escape ladders (for upstairs bedrooms)

Fire extinguishers (one for each level of house, plus kitchen, garage, laundry room, outdoor shed)

N95 or P100 respirator masks

Smoke detectors

CAR SHOPPING LIST (WINTER)

Backpack

Disposable hand warmers

Emergency blanket

Extra clothing

Folding shovel

Food (snack bars)

Jumper cables with portable battery

Mobile phone charger

Sand or cat litter for traction

Snow boots

Water (bottled)

CAR SHOPPING LIST (ALL SEASONS)

Rescue tools (combination glass breaker and seat belt cutter, for driver and passengers)

BIKE SAFETY SHOPPING LIST

Bike helmets (can be used to protect your head in many situations)

Bike light

BLACKOUT/QUARANTINE SHOPPING LIST

Backup generator (natural gas, gasoline, or solar)

Backup ice packs (stored in freezer for emergencies)

Batteries

Candles (battery-operated, or wax)

Cooler for storing freezer or fridge items with ice packs

Emergency/weather radio (hand-cranked)

Flashlights

Headlamp

HVAC filters (7 to 13 MERV)

Matches (for gas pilot lights, candles)

USB charger (hand-cranked)

PANTRY SHOPPING LIST

Baking needs: flours, yeast, sugar

Can opener(s)

Canned soup

Cleaning supplies: mops/sweepers, etc.

Coffee

Condiments

Cooking oils

Dish soap

Dishwasher soap

Grains: rice, oatmeal, pasta, couscous, etc.

Hand soap

Paper towels

Powdered milk

Proteins: nut butters, tinned fish, canned beans, powdered milk

Seasonings: salt, pepper, spices, etc.

Shampoo, razors, toothpaste, and other toiletries

Toilet paper

Tomato sauce, canned vegetables

Trash bags

Vacuum cleaner bags

Vegetable cold storage: potatoes, onions, garlic

Water (bottled)

Utility knife/box cutters

PERSONAL PROTECTIVE EQUIPMENT (PPE) SHOPPING LIST

Alcohol wipes

Face masks

Filters

Gloves

Hand sanitizer

TRAVEL SAFETY SHOPPING LIST

Medical evacuation insurance

Passport

Portable CO detector

Trusted Traveler programs or equivalent in your nation

CYBERSECURITY SHOPPING LIST

Password organizer

Physical login security key

PROTEST PROTECTION SHOPPING LIST

Backpack filled with:

Bike helmet

Ear muffs (sound-deadening)

Extra layers of clothing

Goggles

Pepper spray antidote/ decontamination spray

Water (bottled)

FIRST-AID KIT (SEE COMPLETE LIST ON PAGE 20)

Flexible SAM splints used by doctors, the military, NASA, and other organizations
www.sammedical.store

Pre-made tourniquets used by NATO and U.S. military
http://www.combattourniquet.com/

Pre-made traction splints
https://www.rescue-essentials.com /slishman-traction-splint-compact -sts-c/

Pre-made pressure wraps
https://www.rescue-essentials.com /slishman-pressure-wrap/

Coban and ACE bandages are both products of 3M. Available in all drugstores.

Blister bandages
https://www.compeedusa.com/
https://www.band-aid.com /products/hydro-seal-wound-care

EPA Tick Repellent Finder Tool
https://www.epa.gov/insect -repellents/find-repellent-right -you

EpiPen (requires a prescription)
https://www.epipen.com/

Personal family ID tags:
https://www.roadid.com/

Potassium iodide (available without a prescription)
https://www.cdc.gov/nceh/radiation/emergencies/ki.htm

Tick Encounter Resource Center (Tick Identification Program)
https://tickencounter.org/
https://tickencounter.org/tickspotters/submit_form

BUG-OUT BAG SUPPLIES

Multitools
www.leatherman.com
www.gerbergear.com

Headlamps
https://www.energizer.com/lighting/hands-free-lighting
www.bioliteenergy.com

Knives
https://www.victorinox.com/
https://www.benchmade.com/

Magnesium bars and fire starters
www.uberleben.co
https://firesteel.com/

Mesh network device
https://gotennamesh.com

NOAA weather radios
https://midlandusa.com/

Personal weather systems
https://weatherflow.com/tempest-weather-system/
https://www.acurite.com

Stormproof matches
https://www.ucogear.com/firestarting/matches/

Water purification tools
https://www.lifestraw.com/
https://www.msrgear.com/products/water-treatment

Watches
Marathon supplies watches to the U.S. and Canadian military
https://www.marathonwatch.com

Other reputable makers of rugged watches appropriate for outdoor/survival use
https://www.garmin.com/
https://www.gshock.com/
https://www.suunto.com/
https://www.victorinox.com/us/en/Watches/cms/watches-main
https://www.bertucciwatches.com/
https://www.timex.com/

Bear-proof containers
https://bearvault.com/
https://kodiakcanada.com/
https://ursack.com/

Bear spray and pepper/decontamination sprays
https://www.sabrered.com/
https://www.pepper-spray-store.com/
https://www.mace.com/

Dehydrated meals
https://readywise.com
https://backpackerspantry.com
https://www.mountainhouse.com
https://www.alpineaire.com/us/us
https://www.omeals.com

Pre-packed survival kits
https://unchartedsupplyco.com/

Tactical clothing and gear
https://www.511tactical.com/

Snow boots
https://www.sorel.com

HOME SAFETY

Fire extinguishers, smoke alarms, CO detectors, escape ladders
https://www.firstalert.com/
https://www.shopkidde.com/

Lightning rod installers/contractors
https://lightning.org

Natural gas generators
https://www.generac.com/

Solar generators and panels
https://www.goalzero.com/

CYBERSECURITY

Authenticator apps
Google Authenticator
Authy: https://authy.com/
Microsoft: https://www
.microsoft.com/en-us/account
/authenticator

Credit freezes
https://www.consumer.ftc.gov
/articles/0497-credit-freeze-faqs

Password managers
https://1password.com/
https://www.dashlane.com/
https://bitwarden.com/
https://www.lastpass.com/

Physical login security keys
https://www.yubico.com/

MEDICAL RESOURCES

1918 influenza encyclopedia
https://www.influenzaarchive.org/

Red Cross CPR training
https://www.redcross.org/take-a
-class/cpr

Zombie preparedness:
https://www.cdc.gov/cpr/zombie
/index.htm

TRAVEL

Emergency numbers in countries
outside the United States
https://travel.state.gov/content
/dam/students-abroad/pdfs/911
_ABROAD.pdf

U.S. Trusted Traveler Programs
https://ttp.cbp.dhs.gov/

INDEX

abrasions, treating, 23
abroad
   bioterrorism and traveling, 186
   surviving blackouts, 145–46
active shooter situations, 196–200
adventure packing, 44–48
aerosol weapons, 205
aftershocks, 105
airplane disasters, 214–17
alpha particles, 195
aluminum foil, using with batteries as a fire
   starter, 46
animal attacks, 90–99
anthrax, 181, 182–84, 185
antibiotic ointment, for burns, 26
antihistamine ointment, for insect bites, 28–29
Assisi, Italy, 104
authenticator apps, 211
avalanches, 126–29

babies, considerations for during pandemics,
   151
backup generators, 118, 142
basements, 133
batteries
   for carbon monoxide detectors, 162
   for flashlights, 153
   for smoke alarms, 166
   using with aluminum foil as a fire starter, 46
bear attacks, 94–98
bear spray, 97–98
belts, as tourniquets, 30
bicycles, safety gear for, 232
bikes, safety gear for, 232
bioterrorism, 181–87
black bear, 94, 96
"Black Death," 147–48
blackouts, 140–46, 232
blisters, 26, 125
blizzards, 116–25
bloody wounds, multiple, 29
blunt-force trauma, 134
bombs, 173–78
bubonic plague, 147–48
bug-out bag, 53–54, 55–62, 226–31, 234

buildings, fires in, 165–72
burns, 25–26

camping, wildfires and, 101–2
carbon monoxide (CO), 160–61, 162–63
carbon monoxide detectors, 140–41, 162
cardiac arrest/heart attack, 36–37, 40–41,
   224–25
cardiopulmonary resuscitation (CPR), 32–34,
   36, 138–39
cars. See vehicles
cash on hand, 154
castor bean plant, 79
cat attacks, 92–93
cat-scratch fever, 93
chemical agents, 188–92
chlorine gas, 189–90
chlorine/chlorine tablets, 75, 163–64
choking, 35
cities, surviving blackouts in, 144–45
civil unrest, 201–5, 233
cleaning supplies, during pandemics, 155
clear air turbulence (CTA), 216
CN compounds, 205
cold, for treating burns, 25
compasses, 47, 66
concussion, 24–25
condoms, 46
COVID-19 coronavirus, 147–50
CPAPs, 151
crank radio, 152–53
crowd-control weapons, 204
"CS" tear gas, 203, 205
cuts, treating, 23
cyberattacks/cybersecurity, 206–13, 233,
   235

daffodil, 79
day hike packing list, 45
dehydration, 105, 120
diabetics, grand mal seizures in, 39, 42
digital thermometer, 151
direct pressure, for wounds, 23 29
dirty bombs, 193–95
dog attacks, 92–93

drowning, 38–39
duct tape, 46, 153

E. coli, 187
earthquakes, 104–6
elders, considerations for during pandemics,
   151
elevator accidents, 179–80
emergencies
   burns, 25–26
   cardiac arrest/heart attack, 36–37
   cardiopulmonary resuscitation (CPR),
      32–34
   choking, 35
   concussion, 24–25
   cuts/abrasions, 23
   drowning, 38–39
   eye burns/splashes, 32
   impalement, 31
   insect bites, 28–29
   insect stings, 26–27
   multiple bloody wounds, 29
   seizures, 39–42
   stroke, 42–43
   tourniquets, 29–30
   treating basic, 21–43
emergency hand-cranked USB charger, 142
epilepsy, 39, 42
EpiPen, 26, 27
escape ladder, 170
escape plan, 171
essential workers, 156
evacuation
   about, 51–52
   dirty bombs and, 195
   from hotels in fires, 171–72
   mudslides and, 114
   wildfires and, 101, 102–3
eye burns/splashes, 32

Face, Arms, Speech, Time (FAST) acronym,
   42–43
face masks, 156
ferning, 136
ferro rods, 64–65
fire extinguishers, 167–68
fire 'hygiene,' 169
fire safety plan, 168–69
fire starter, 63–65
fires, 165–72, 232
first-aid kit, 19–20, 151, 222, 233–34
flash-bangs, 204
flashlights, 153

flints, 64–65
floods, 107–10
flu, 181
foxglove, 80
fridges/freezers, 17, 142–43
frostbite, 118, 121–22, 124–25

gamma particles, 195
garage doors, 17
garages, carbon monoxide and, 162
gas leaks, 17, 106, 160–64
gasoline-powered generator, 142
gauze, 23, 26
gear, where to buy, 232–35
generators, backup, 118, 142
Good Samaritan laws, 34
grand mal seizure, 39
grizzly bear, 97
grounded, being, 137

H1N1 virus, 147
hand sanitizer, 46, 155
hand-cranked emergency radio, 142
heat exhaustion, 143
Heimlich maneuver, 35
home fires, 165–72
home safety
   about, 14–16
   action steps for, 16–18
   adventure packing, 44–48
   bug-out bag, 53–54, 55–62
   chemical agents, 191
   drinking dirty water, 73–75
   evacuation, 51–52
   family rendezvous plan, 84–86
   fires, 169–72
   first-aid kit, 19–20
   HVAC system, 185
   Rule of Three, 49–50
   shopping list for, 234
   surviving blackouts, 141–43
   toxic plants, 76–83
   treating basic emergencies, 21–43
   using fire starter tools, 63–65
   using wristwatches for directions, 66–72
   zombies, 87–88
hospitals, for gas leaks, 163
human-caused disasters
   airplane disasters, 214–17
   bioterrorism, 181–87
   chemical agents, 188–92
   civil unrest, 201–5
   cyberattacks, 206–13

human-caused disasters (*continued*)
  dirty bombs, 193–95
  elevator accidents, 179–80
  gas leaks, 160–64
  home/building fires, 165–72
  shooters/stabbers, 196–200
  taking a pulse, 218–19
  terrorist activities, 173–78
hurricanes, 107, 108, 140
HVAC system, 185
hydrogen peroxide, for cuts/abrasions, 23
hyperthermia, 143
hypothermia, 118, 121–22, 123

ice, 24, 26
identity theft, 211–12
impalement, 31, 134
infants/small children, CPR on, 34
insect bites, 28–29
insect stings, 26–27
iodine tablets, 75
isolation plan, 157

Jimson weed, 76

Khan, Alli S., 87

larkspur, 80
LifeStraws, 15
lightning, 135–39
lost, getting, 47–48, 66–72

magnesium bars, 64–65
mechanical watch, 71–72
medical data, protecting, 213
medical resources, 235
medical treatment, deferring, 150
monkshood, 81
Morse code, 48, 106, 223
mosquitoes, 28
mountain laurel, 81
mudslides, 111–15
mushrooms, 77

N95 respirators, 156, 185
National Fire Danger Rating System (NFDRS),
  102
National Oceanic and Atmospheric
  Administration (NOAA), 152
natural gas, 160, 161
natural perils
  animal attacks, 90–99
  avalanches, 126–29

blackouts, 140–46
blizzards, 116–25
earthquakes, 104–6
floods, 107–10
lightning, 135–39
mudslides, 111–15
pandemics, 147–58
tornadoes, 130–34
wildfire, 100–103
non-perishables, 153–54
notes, leaving when hiking, 47

OC (oleoresin capsicum), 204
office buildings, surviving fires in, 171–72
oleander, 79, 81
oximeter, 151
oxygen, fires and, 169

packing, adventure, 44–48
pain, from frostbite, 125
pandemics, 147–58
pantry, shopping list for, 233
P.A.S.S. acronym, 167–68
password protection, 210–11
patient portal, 150–51
pepper spray, 97
perishables, 153–54
personal data, protecting, 212
Personal Protective Equipment (PPE), 155–57,
  233
plague, 94
plants, toxic, 76–83
poison ivy/oak/sumac, 76, 77–78
pokeweed, 82
potassium iodide, 195
prescriptions, 150–51
Primary, Alternate, Contingency, Emergency
  (PACE) acronym, 84–85
protests. See civil unrest
pulse, taking a, 218–19
purifying water, 74–75

quartz watch, 70–71

rabid animals, 91–92
rabies injections, 91–92
radiation, 195
radium-226, 193–94
ransomware attacks, 207
rat attacks, 93–94
rat-bite fever (RBF), 93–94
rhabdomyolysis, 105
ricin, 188

ring vaccinations, 185
ROAD iD-brand silicone bracelets, 15, 18
Rule of Three, 49–50, 116–17

sarin, 190
scams, loved ones and, 212–13
seizures, 39–42, 40–41, 224–25
shelters
    during lightning, 137
    surviving blizzards in, 117–19
shooters/stabbers, 196–200
sleep, concussions and, 24–25
smallpox, 181, 182–83
smartphone apps, 17–18
smoke alarms, 166
snake bites, 98–99
solar generator, 142
solar power emergency radio, 142
spam calls, 209–10
spam folders, in email, 208
specialized gases, 163–64
spit drool test, 13, 129
sporting/recreational equipment, 18
stick-and-shadow method, for finding your
    direction, 67–69
stinging nettle, 82
stroke, 40–41, 42–43, 224–25
"stroke-like" symptoms, for concussions, 24
stun grenades, 204
surge suppressors, lightning and, 138
Survival, Evasion, Resistance, Escape (SERE)
    training, 9–10
survival knife, 153

tampons, as kindling, 46
tear gas, 201, 203, 205
telehealth, 151
terrorist activities, 173–78
ticks, 28
toilet paper, 156
toiletries, during pandemics, 155–56
tornadoes, 130–34
tourniquets, 29–30
toxic plants, 76–83
toxins, 166, 188–92

trampolines, 14
travel
    in floodwaters, 108
    during pandemics, 157–58
    resources for, 235
    shopping list for safety, 233
trees, lightning and, 138
Trusted Traveler program, 158
two-factor authentication (2FA), 210–11

U.S. Centers for Disease Control and
    Prevention, 87–88
The U.S. Air Force Pocket Survival Handbook,
    10–11

vaccinations, 185
vehicles
    batteries for, 17
    breaking windows in, 109–10
    chemical agents in, 191
    driving through standing water, 109
    gas for, 17
    keyless ignition for, 162
    during lightning, 137
    in mudslides, 112
    shopping list for safety gear, 232
    surviving blizzards in, 119–121
    during tornadoes, 134
ventricular fibrillation, 139
Venturi effect, 134
vespid allergies, 26–27

Waldo Canyon, Colorado, 100–101
water
    drinking dirty, 73–75
    lightning and, 137
    purifying, 74–75
water hemlock, 77, 83
wildfires, 100–103
windlass, in tourniquets, 29–30
windows, breaking in vehicles, 109–10
wireless emergency weather alerts, 133
wristwatches, using for directions, 66–72

zombies, 87–88

ABOUT THE AUTHOR

Dr. John Torres is currently the NBC News, Today Show and MSNBC senior medical correspondent. An emergency room physician who has extensive international experience in the field—and on the front lines—Dr. Torres brings a vast breadth of knowledge to NBC's medical coverage. His reporting covers a wide range of health-related issues across all of NBC's broadcast, cable, and digital platforms.

Dr. Torres is a self-described "Air Force Brat," having grown up traveling and living around the world with his family and Air Force father. Along with learning many life lessons from his dad, Torres also followed in his father's footsteps and joined the Air Force. Dr. Torres graduated from the U.S. Air Force Academy and served for decades in the U.S. Air Force, Reserves, and Air National Guard. During his active duty years, Dr. Torres served as a pilot, then attended medical school and returned to the service of his country as a physician and flight surgeon. He is a proud veteran of the United States Air Force where he served for 8 years, including a tour of duty in Iraq in 2004 with the Air National Guard. He later served an additional 24 years in the Air National Guard and Air Force Reserves, before retiring as a Colonel. An accomplished pilot and physician, Dr. Torres has contributed to rescue efforts out of the South Pole and, in the wake of Hurricanes Katrina and Rita, set up medical care units and led rescue missions.

Utilizing his combined medical and military experience, Dr. Torres helped develop and implement training courses for NATO Special Forces soldiers to help ensure a high level of consistency across a variety of nations, languages, and cultures. He continues to teach NATO Special Forces a variety of skills, including tactical combat casualty care, combat simulations, and medical leadership. Additionally, throughout his career, Dr. Torres has made numerous humanitarian trips to Central and South America, providing medical care to children in need.

Dr. Torres joined NBC's medical unit from 9News, the NBC station in Denver, Colorado, where he answered viewers' medical questions and reported on breaking health stories and their impact on communities, all while running a successful clinical practice as an emergency room physician. A graduate of the University of New Mexico School of Medicine, Dr. Torres continues to practice medicine, teach NATO personnel, U.S. medical students and residents, and provide science-based information to viewers across the nation and world in a way that helps viewers live their lives in a happy and healthy way. Dr. Torres lives with his wife in Colorado and New York, and has two adult children.